SpringerBriefs in Energy

Energy Analysis

More information about this series at http://www.springer.com/series/10041

Nafeez Mosaddeq Ahmed

Failing States, Collapsing Systems

BioPhysical Triggers of Political Violence

Nafeez Mosaddeq Ahmed
Global Sustainability Institute
Anglia Ruskin University
Cambridge, UK

ISSN 2191-5520 ISSN 2191-5539 (electronic)
SpringerBriefs in Energy
ISBN 978-3-319-47814-2 ISBN 978-3-319-47816-6 (eBook)
DOI 10.1007/978-3-319-47816-6

Library of Congress Control Number: 2016955192

Printed on acid-free paper

This Springer imprint is published by Springer Nature
The registered company is Springer International Publishing AG
The registered company address is: Gewerbestrasse 11, 6330 Cham, Switzerland

Contents

Chapter 1
Introduction

Since the 2008 financial crash, the world has witnessed an unprecedented outbreak of social protest in every major continent. Beginning with the birth of the Occupy movement in the US and Western Europe, and the Arab Spring, the eruption of civil unrest has continued to wreak havoc unpredictably from Greece to Ukraine, from China to Thailand, from Brazil to Turkey, and beyond. In some regions, civil unrest has coalesced into the collapse of incumbent governments or even the eruption of a prolonged state of internecine warfare, as is happening in Iraq-Syria and Ukraine-Crimea. To what extent is this apparent heightening of geopolitical instability new?

Increasing public dissatisfaction with government is correlated with continued government difficulties in meeting public expectations. Yet while policymakers and media observers have raced to keep up with events, they have largely missed the deeper causes of this new age of unrest—the end of the age of cheap fossil fuels, and its multiplying consequences for economic growth, industrial food production, and the Earth's climate stability.

Contrary to widely reported claims across mainstream media of a new era of prosperity heralded by the US-led shale oil and gas boom, the proliferation of contemporary climate, food and economic crises have at their root a single common denominator: the fundamental and permanent disruption in the energy basis of industrial civilization.

This inevitable energy transition away from high quality fossil fuels to lower quality, more expensive energy forms—which will be completed well before the close of this century, and quite possibly much earlier—will force a paradigm shift in the organization of civilization. The twenty-first century, in this context, is a pivotal one for humanity as industrial civilization pivots through a process of systemic transition, driven by the complex interplay between human societies and biophysical realities.

Yet for this shift to result in a viable new way of life will require a fundamental epistemological shift recognizing humanity's embeddedness in the natural world. This, in turn, cannot be achieved without breaking the stranglehold of conventional

N.M. Ahmed, *Failing States, Collapsing Systems*, SpringerBriefs in Energy,
DOI 10.1007/978-3-319-47816-6_1

models achieved through the hegemony of establishment narratives—dominated by fossil fuel interests and the banality of the mainstream media news cycle.

The central thesis of this study is that the escalation of social protest and political instability around the world is causally related to the unstoppable thermodynamics of global hydrocarbon energy decline and its interconnected environmental and economic consequences. It offers, in this sense, a biophysical approach to international relations, and argues that geopolitics remains fundamentally embedded in biophysical processes. This is not to reduce geopolitics to the biophysical—far from it—but to recognize that the dynamics of the geopolitical cannot be dislocated from the dynamics of the biophysical, and that biophysical processes are increasingly driving geopolitical instability to a degree unrecognized by policymakers, the media, as well as social and natural scientists.

Hydrocarbon energy decline can be understood as consisting of the following two intertwined processes: the inexorable reduction in industrial civilization's production of net energy from hydrocarbon sources (fossil fuels) over the last decades; the acceleration of hydrocarbon energy production to attempt to make-up for this decline and sustain economic growth.

This process has in turn had two major consequences, namely: climate change and the corresponding destabilization of the Earth System due to the increasing quantity of greenhouse gas emissions due to hydrocarbon energy dependence; and the permanent slowdown of global economic growth due to the increasing costs of energy production relative to GDP. Climate and economic crises are, in turn, acting as amplifying feedbacks on the process of hydrocarbon energy decline, and in themselves are acting synergistically to undermine global industrial food production while simultaneously impinging on socio-political stability and human well-being.

While conventional policy and media approaches to socio-political instability tend to focus almost purely on 'surface' social symptoms—geopolitical competition, political corruption, economic mismanagement, ideological extremism, and so on—the deeper biophysical systemic drivers of instability are largely ignored or misunderstood. As such, missing from the vast bulk of conventional wisdom on escalating socio-political instability around the world is the crucial recognition of its central cause in a systemic process of hydrocarbon energy decline and concomitant civilizational transition toward an inevitable post-carbon future.

Currently, climate change is rightly and consensually recognized by the scientific community, and has been accepted at least in principle by policymakers as a reality requiring an urgent collective response from human societies. However, despite growing recognition of the interconnected nature of these crises—illustrated through concepts such as the food-water-energy nexus—there remains a fundamental failure in the conceptualization of their interconnected nature in terms of the relationship between human societies and the biophysical environment, and relatedly, the relationship between human polities and the biophysical environment.

This root failure of *conceptualization* is perhaps the most significant factor focusing on the role of human agency in driving the current convergence of global crises. The failure is compounded by the necessarily compartmentalized nature of scientific specialization, which has produced a vast volume of information, but little in the way of epistemological mechanisms to integrate that into knowledge across disciplinary boundaries. It is further compounded by the transmission of incorrect conceptual diagnoses of global crises through the global mainstream media. The perpetual transmission of false and inaccurate knowledge on the origins and dynamic of global crises has created a situation in which as such crises accelerate, the human species as a whole is disempowered from being able to correctly understand these crises and their symptoms, and thus unable to solve them.

Yet to diagnose the intensifying perfect storm of climate, energy and economic crises requires a fundamental reconceptualization of their true nature as symptoms of an overall civilizational system which, increasingly, cannot be sustained by the biophysical environment.

This study offers an empirically-grounded social scientific theoretical framework for developing a holistic approach to this perfect storm through the lens of what this author has called the 'Crisis of Civilization' (as opposed to a 'clash of civilizations'). My approach is not to create yet another new statistical model to add to the plethora of models that exist, but to strike at a deeper lacuna within the discipline of international relations—to create the beginnings of an accurate, integrated transdisciplinary theoretical basis for such modelling, derived from a holistic analysis of the relevant empirical data.

The study is divided into two main sections. The first consists of a general framework of the broad crises that this author considers to be integral to the biophysical processes driving geopolitical instability today. The second consists of a series of case studies which provide specific empirical data supporting and building on this general framework to test how and whether they are indeed acting out on local, national and regional levels as is my hypothesis. This opening section begins by building on this author's previous work on the 'Crisis of Civilization' as an overarching analytical framework for the integrated examination of global climate, energy, food, economic and socio-political crises (Ahmed 2010, 2011). This is achieved by establishing the inherent systemic interconnections between these crises on a global macro-level scale. The monograph then proceeds to explore how the general framework of biophysical factors aka the 'Crisis of Civilization' plays out at a micro-level within specific countries in key regions—the Middle East, Africa, Europe, South Asia, and North America. This examination takes us to the crux of my argument and provides specific evidence that the biophysical processes discussed more generally in the opening are, in fact, already having concrete geopolitical impacts accelerating the destabilization of human societies across the world, in a manner that can now be detected through a holistic and transdisciplinary empirical analysis. This not only provides surprising empirical vindication for our hypothesis that biophysical processes are playing an integral causal role in the intensification of political and geopolitical disruption on a global scale, it also provides us a basis to explore some tentative business-as-usual (BAU) forecasts.

This permits further exploration of the intersection between the thermodynamics of escalating hydrocarbon energy decline and the accelerating disruption of global industrial civilization. As prevailing social, political and economic structures become increasingly dysfunctional against the strain of hydrocarbon energy decline, the resulting rupture manifests in an increasing frequency of social protest and violent conflict.

Part of this study, then, identifies how conventional governmental, industry and media narratives of these crises for the most part fail to accurately understand them, not just due to a lack of a holistic-systemic frameworks for examining these crises as interdependent—but due to a fundamental epistemological failure that has allowed mythological 'theories' of human progress in the form of neoclassical and neoliberal economics to become entrenched as the dominant cognitive paradigm.

The most powerful hegemonic component of this ideological capture of human collective cognition occurs through the global institutions associated with the mainstream media. The principal problem here is a highly compromised ownership and editorial structure that ties media outlets to the very prevailing structures of fossil fuel-centric power complicit in global crisis acceleration. The preponderance of fossil fuel-centric interests in conventional media ownership has led to consistently inaccurate reporting on energy issues, and their relationships with economic, food and climate crises, as well as specific conflicts.

Yet to some extent, and compounding the insular ideological approach of powerful government, industry and media institutions, there has been a similar failure from amongst experts in different fields of these crises, who have been unable to develop theoretical, conceptual and empirical frameworks to view their specialized data in its inherent interconnections with data from other fields. In other words, a lack of generalized systems training in our schools. Due to this problem, we are beginning to grasp only recently the extent to which geopolitical ruptures that overwhelm of the news of the day have been exacerbated by a convergence of crises studied largely separately in these disparate fields. There is, therefore, little understanding of how energy and resource depletion tangibly impact the political economies of different societies, and how these processes interact with the local impacts of global processes like climate change.

This has led to a knowledge deficit—specifically, *a whole systems knowledge deficit* comprising a paucity of reliable, actionable knowledge in the mainstream, exacerbating a sense of public apathy and confusion, and cementing a policymaking impasse among political leaders who remain subject to a fatal combination of intensive fossil fuel lobbying and media misinformation.

Among the most critical solutions to the 'Crisis of Civilization', then, is a concerted grassroots mobilization to rectify the *whole systems knowledge deficit*. This could be achieved in many different ways—whether through responsible journalism or more informed policy formulation based on the effective communication of interdisciplinary scientific research—but the end goal is the same: mass public education with a view to catalyze social action that is systemically transformative. Without addressing the knowledge deficit, the self-reinforcing cycle of amplifying crisis feedbacks cannot be overturned.

It is hoped that this study can begin contributing to addressing the whole systems knowledge deficit by firstly, establishing a scientifically-grounded systems theory framework for integrating data from different fields for the study of international politics; secondly, beginning the process of recognizing major geopolitical ruptures in the context of systemic crises driven by biophysical processes; thirdly, outlining the basis for a major, urgent new transdisciplinary research program bringing together the natural and social sciences to develop a holistic theoretical-empirical model of global crisis convergence; which in turn can pave the way for a fourth major, urgent new transdisciplinary action-research program on mitigating the impact of global crisis convergence, while transitioning human civilization to new more viable political economic structures that subsist in parity with their biophysical contexts.

Chapter 2
The Crisis of Civilization as an Analytical Framework

2.1 The Human-Environment System as a Complex Adaptive System

The idea of a 'Crisis of Civilization' pivots around the goal of understanding human activity as a whole. It is premised on the fact that as a single biological species, human beings share common individual and social characteristics through which they interact with each other, with other species, and with the biophysical environment.

Global civilization constitutes the full mechanism of social organization by which this nexus of activities and interactions operates.

I use a 'Crisis of Civilization' framework to examine multiple, seemingly disparate global and local crises. This does not obviate the specific and distinctive dynamics of those crises, but permits examination of how these crises interrelate with one another in the context of the overarching global system of which they are part.

The theorization of human civilization as a "complex adaptive system" derives from the application of complex systems theory as developed in relation to biological systems and ecosystems (Kauffmann; Dyke; Homer-Dixon; Diamond). A rich and dense literature demonstrates that complex systems are found across the natural sciences in physics, chemistry, and biology (Ross and Arkin 2009), as well as in ecology (May et al. 2008) and economics (Farmer et al. 2012).

A system exists whenever a plurality of entities subsists in which each entity functions in some sort of relationship with the others. A *complex* system exists when the relations between these parts leads the system as a whole to display emergent properties and behavior which cannot be reduced solely to the nature of its different parts and their relationships. Those emergent properties can be codified as overarching rules that characterize the system's structure as a whole. In some cases this can be done mathematically, although this is a less useful approach when examining human societies.

N.M. Ahmed, *Failing States, Collapsing Systems*, SpringerBriefs in Energy,
DOI 10.1007/978-3-319-47816-6_2

A *complex adaptive* system exists when the system as a whole is able to adapt—to generate a collective shift in its internal behavior in order to survive. Thus, while the relations between parts of a system generates the emergent structures that comprise the system as a whole, those relationships are, in turn, restrained and enabled by those wider structures. This circular relationship is integral to the system's capacity to adapt to new environmental conditions. In time this is done by evolution; more immediately this can be done by behavioral changes or species shifts.

Equally, due to the nested and interconnected nature of the components of a complex adaptive system, small perturbations in one part can have ramifying effects on other parts of the system, depending on how they are connected. This sort of internal positive feedback process means that the overall structure of a system can be greatly impacted by seemingly random occurrences—those structures can either be reinforced or undermined by these internal feedback processes.

When such internal feedback processes reach certain thresholds, or 'tipping points', they can induce fundamental re-ordering of key structures in the system as a whole—the convergence of multiple tipping points, in turn, can generate a system-wide adaptive cycle of re-structuring, a 'phase shift', through which the system undergoes a transition to a new equilibrium (Holling 2001).

The human-environment system is complex and adaptive because it represents a historically evolving civilizational form comprised of a vast interlocking array of nested sub-systems, including some from the earth's geology, resources, oceans, and atmosphere; multiple living and non-living ecosystems across these domains; and human systems, comprised of psychological, cultural and ideological fields, relations of production and associated modes of energy extraction, technological and economic systems, and political structures.

Thus, the 'Crisis of Civilization' framework is a systems approach that attempts to analyze the complex interrelationships between multiple global crises and human activities as a whole, thus understanding them not simply as discrete crises and activities in themselves, but as component factors of a wider global human-environmental system with its own emergent properties and behaviors.

This approach recognizes that each of these crises pertains to a specific sub-system in itself, with distinctive features and patterns of behavior, but equally recognizes that each of these sub-systems do not exist in isolation. Rather, their mutual interrelationship generates emergent patterns characterizing the system as a whole. Those emergent structural features, in turn, exert causal regulatory effects that shape, enable and constrain the behaviors of the sub-systems.

A systems approach thus views particular crises and associated human activities as discrete sub-systems which are, nevertheless, inherently interconnected as sub-systems in an emergent human-environmental system, captured through the concept of a world-scale human civilization. It is in this respect that the 'Crisis of Civilization' as an analytical framework is able to systemically locate multiple crises as interconnected features of a wider world-scale crisis in human civilization as an emergent macro-structure. By integrating detailed trans-disciplinary examination of crisis sub-systems with analysis of their systemic interconnections within the world-scale human-environment system, a much clearer picture of the precise drivers, dynamics and potential trajectories of these crises is possible. This permits discernment of a birds-eye perspective of overall civilizational structures and their emergent direction.

Examining human civilization as a complex adaptive system, therefore, permits multiple global crises to be understood through the lens of a range of powerful concepts with solid empirical basis in the biophysical sciences—the thresholds and tipping points of feedback processes; how interconnections between different crises can generate amplifying feedbacks with the potential to accelerate the breaching of tipping points; the extent to which different crises can be seen as properly systemic—that is, related fundamentally to the key global structures integral to the prevailing dynamic of human civilization; and how these crises relate to the system's adaptive capacity, in particular, whether they are generating a major 'phase change' in the system itself.

In particular, this allows analysis of human civilization to return to a scientific framework defined by the thermodynamics of the fossil fuel system, and the evolution and adaptation of species, bringing in critical insights from the physical and natural sciences that can inform the development of robust historical and sociological theories.

2.2 The Energy Metabolism of Human Civilization

Applications of complex systems approaches to social, environmental and economic phenomena have largely neglected the most fundamental factor in the evolution and adaptation of complex systems: energy metabolism.

Extensive research in the biological and ecological sciences demonstrates that an organism's relation to the environment is mediated fundamentally through the mode and manner by which it extracts energy from the environment, to maintain and improve its distance from thermodynamic equilibrium. Living systems extract free energy from the sun, store it, and use it. Further, they can reproduce as well as collect, process, and exchange information in order to control and direct energy and matter they receive from their environments (Terzis and Arp 2011; Hall et al. 1992).

According to the Second Law of Thermodynamics, physical systems display a tendency to dissipate energy and thus transition from states of order to increasing disorder. Therefore, physicist Erwin Schrödinger defined a living system as an embodiment of "negative entropy" as they "extract order" from their environments to survive, adapt and evolve. (Schrodinger 1944)

A living system or organism is thus defined by its ability to store energy under energy flow, before dissipation. It develops, maintains and reproduces, or renews, itself by mobilizing material and energy extracted from the environment, which is stored internally through cyclic non-dissipative processes coupled to irreversible dissipative processes. This permits the organism to survive precisely through the consumption and ordering of energy within systemic biological processes organized through genetic information protocols. The capacity to extract, store and mobilize stored energy is therefore integral to a reproducing life cycle. Eventually, of course, the energy must be irreversibly dissipated as required by the Second Law. But the increasing complexity of a living system is related directly to its capacity to extract, store and mobilize stored energy, and to thereby stave off the thermodynamic dissipation of energy. (Ho 1999)

Organisms which successfully adapt to changing or challenging environmental conditions do so through the superior processing of information about those external conditions through genetic modification, reflecting increased efficiencies in energy extraction, storage and mobilization in relationship with the environment. (Schneider and Kay 1994)

The thermodynamics of living systems applies, of course, not just to any single individual organism, but simultaneously to collections of organisms inhabiting specific environments. While human beings are the most advanced—that is, complex—biological organisms known to science, human civilization constitutes a complex adaptive system which has been able to maximize energy extraction, storage and mobilization from its environment far more efficiently and powerfully than ever before. The astonishing complexity of human civilization is related directly to its capacity to harness energy from the environment through numerous sub-systemic processes of social organization, thus maintaining increasing distance from thermodynamic equilibrium (Odum 1994).

This framework allows for a more complete empirically-grounded theorization of what the contemporary escalation of global environmental and economic crises entails for the current trajectory of human civilization. Over its historical evolution, human civilization has demonstrated a relationship with its environment involving escalating energy use and energy dissipation, with wide-ranging consequences for the stability of the global human-environment system.

Social power is an organic constitution grounded in an exploitative relationship with nature by which energy is extracted from natural resources, transformed into a commodity (through production) and eventually consumed. Energy is thus the very condition of production—but to examine the fluctuating relationship between the two requires the recognition of social power through *property*: that is, the way access to land, resources and technology to enable energy production is mediated through property rights, which in turn are related to configurations of class (Wood 1981; Aston et al. 1987; Rioux and Dufour 2008).

It is therefore necessary, in examining the energy trajectory of human civilization as a whole, to investigate inequalities in social power and class in the context of differentiated access to land, resources and technology between various human groups, and how this relates to the thermodynamics of energy as applied to human society as a complex adaptive system. This will enable us to properly grasp the processes of extraction, transformation and consumption of energy through labor, and varying relationships between society, labor, technology and natural resources, that are integral to diagnosing human civilization's current predicament (Foster et al. 2010; Hall and Klitgaard 2012).

The historical development of human civilization illustrates an accelerating trend in global net energy production driven by a series of increasingly sophisticated technological breakthroughs, each linked to fundamental shifts in the human-environment relations and corresponding socio-political and economic systems of organization. These civilizational phase-shifts toward more complex forms can be conceptualized in multiple overlapping ways.

These phase-shifts have encompassed fundamental transitions in the energy metabolism of human societies—in terms of both the types of energy extracted, and the relations of production by which this energy is extracted, stored and mobilized in society through the creation of goods and services. These energy sources include our own muscle and that of animals, as well as wood, wind, water, coal, oil, and nuclear power (LePoire et al. 2015). The relations of production accompanying these phase shifts have included the following social-property relations: hunter-gatherer, nomadic, pastoral, agrarian, feudalism, slavery, agrarian capitalism, industrial capitalism, and neoliberal finance capitalism, which is rapidly moving to a new phase of late capitalism predominated by information technology and artificial intelligence (Ahmed 2009).

It is also important to note that each new phase-shift does not necessarily constitute a clean break with previous shifts, but in the course of increasing complexity often builds on or incorporates older structures within a new, wider structural context. One useful way of understanding this process in an evolutionary fashion is through Arthur Koestler's concept of nested self-organizing hierarchical systems which successively incorporate less complex systems to create higher scales of overarching complexity (Pichler 1999).

While the latest phase shift of neoliberal finance capitalism has been able to generate an unprecedented level of wealth within the system, it has simultaneously developed an unprecedented degree of global inequality between the core—consisting of a transnational nexus of class power centred in the former G8—dominating the world's productive resources including energy, raw materials, military and information technology; and the periphery—whose countries remains largely subordinated to the global structures institutionalized by the core (Ahmed 2009; Tainter 1990).

2.3 The Physics of System Failure

Today, human civilization under late capitalism maintains its increasing distance from thermodynamic equilibrium via the throughput of vast quantities of increasingly depleted fossil fuel reserves, along with other finite and increasingly scarce resources such as metal ores, radionucleotides, rare earth elements, phosphate fertilizer, arable land, and fresh water (Nekola et al. 2013).

One indicator of the system's growing complexity today is the measure of material throughput, or economic growth—Gross Domestic Product (GDP). Under capitalist social-property relations, GDP must continuously increase through the maximization of private sector profits, simply for businesses to survive in the competitive marketplace and for the economy to maintain its ability to meet the consumption requirements of a growing population. However, as the complexity of human civilization has advanced, the continual growth in material throughput is correlated with an escalating rate of depletion of energy and raw materials, as well as an acceleration in the dissipation of energy through intensifying greenhouse gas emissions.

Robust scientific assessments now demonstrate that the continuation of those biophysical processes of environmental degradation in a business-as-usual scenario will, before the end of the twenty-first century, fundamentally undermine the biophysical basis of human civilization in its current mode of material organization and structural complexity. Further, the uncontrolled energy releases generated by these biophysical processes are manifested in climate change, extreme weather events, and natural disasters (Earth System Disruption); and drives geopolitical competition, social unrest, and violent conflict (Human System Destabilization).

These manifestations of dissipative energy release can be seen as distinctive feedback processes resulting from human civilization's accelerating exploitation of fossil fuel energy sources within the context of the biophysical limits of the environment. In turn, these two strands of systemic feedbacks—Earth System Disruption (ESD) and Human System Destabilization (HSD)—are occurring within a single, overarching human-environment system, and thus are already inherently interconnected, therefore feeding back into each other.

This mutual feedback process creates an amplifying global systemic feedback in which: (1) ESD drives HSD, which in turn generates 'security' issues perceived through the lens of 'threat' and 'risk' analysis; (2) this invites traditional securitized human responses that focus on the expansion of existing military, political and economic power to stabilize existing structures of authority and advance prevailing mechanisms of energy extraction and mobilization; (3) the entrenchment and expansion of existing structures undermines human civilization's capacity to pursue structural modifications to ameliorate, mitigate or prevent ESD, thus intensifying ESD; (4) the feedback process continues as ESD drives further HSD.

The trajectory of this amplifying global systemic feedback, carried to its logical conclusion and assuming no intervening shift, is simply the protracted, cascading collapse of human civilization in its current form toward increasingly less complex, and therefore less resource-intensive configurations, corresponding to available resources and constrained within the environmental limits imposed by accelerating climate change (Tainter 1990).

Within this amplifying global systemic feedback, one fundamental obstacle to the systemic restructuring required to avert this outcome is knowledge access, distribution, and processing. In much the same way that an integral factor in an organism's capacity to adapt to changing environmental conditions is its genetic ability to absorb environmental information and process it through genetic modification that can result in new adaptive biological configurations, human civilization must be capable of absorbing and processing accurate information about the human-environment system, and converting this into actionable knowledge, in order to be empowered to enact the key structural modifications capable of effecting a phase-shift to a more stable adaptive configuration in relation to the Earth System.

The difference here, of course, is that while evolutionary biological genetic modification is a question of random mutations, human civilization consists of a collection of conscious agents who can make deliberative decisions on the basis of the information available to them, which must be integrated into knowledge that is capable of informing adaptive behaviors. This raises the question of a pivotal system-wide structural deficiency in the knowledge processing capacity of human civilization. In short, inaccurate, misleading or partial knowledge bears a particularly central role in cognitive failures pertaining to the most powerful prevailing human political, economic and cultural structures, which is inhibiting the adaptive structural transformation urgently required to avert collapse. The most obvious locus of this global systemic information deficit is, of course, the global media system—or perhaps more accurately, the Global Media-Industrial Complex (GMIC), and related organs of communication and transnational information dissemination. The GMIC, in effect, currently operates as the information-knowledge architecture of human civilization.

The implications of this analysis are stark: scientific data demonstrates that the rapid convergence of multiple global crisis in coming years and decades is pushing a vast array of interconnected sub-systems toward a threshold of simultaneous tipping points. From a complex adaptive systems perspective, this feedback threshold signifies a global system that is on the brink, if not in the midst, of a fundamental phase-shift to a new structural configuration.

However, the evolutionary context of this process suggests that the nature and outcome of this global civilizational phase-shift will determine the ultimate fate of civilization. Rapidly changing environmental conditions and the escalating breach of biophysical limits are compelling human civilization to either adapt through fundamental structural reorganization, or face a cascading and uncontrolled series of structural regressions.

In the following sections, the theoretical architecture laid out here will be elaborated more precisely with respect to empirical data across the sub-systems of energy, mineral resources, climate, the economy, food production, and civil unrest.

Chapter 3
Net Energy Decline

The rate of growth of human civilization's global net energy production for the first time in history began to slow down since the end of the twentieth century (King 2015; King et al. 2015a, b). Global net energy production may have already reached, or else is rapidly approaching, a peak as the rate of growth in energy production declines, and as the quality of traditional mineral sources of energy also declines (Fig. 3.1).

The prospect of the peak and decline of net energy production relates to the concept of Energy Return on Investment (EROI), a ratio which calculates resource quality by comparing the quantity of energy extracted with the amount of energy inputted to extract that energy. High resource quality is attained the larger the value of energy output, relative to the energy input. The lower the value of the energy extracted, and the more energy inputted to extract it, the lower the resource quality (Hall and Klitgaard 2012).

Since the mid-20th century, the EROI of hydrocarbon energy sources has experienced an overall decline, driven largely by two biophysical factors: the depletion of high quality resources and therefore the increasing reversion to hydrocarbon sources whose exploitation produces lower quantities of energy; thus the rise in the energy cost required to extract these lower quantities of energy.

Between 1960 and 1980, the world average value of EROI declined by more than half from 35 to 15 (Castillo-Mussot et al. 2016; Hall et al. 2014; Hall and Klitgaard 2012; Hallock et al. 2014). During this very period, although actual fossil fuel production continued to increase, since 1950 the rate of production has been declining (Fig. 3.2).

Among the key drivers of this decline in resource quality despite increasing production is the growing shift toward unconventional fossil fuels, which are more expensive and difficult to produce, and whose energy value is lower than that for conventional oil. Since 2005, the rate of increase of conventional oil production has dramatically slowed, to the point that it appears to now be on an undulating plateau that has been unable to exceed a ceiling of around 75 million barrels per day (mbd) (Murphy and Hall 2011). Meanwhile various "petroleum liquids", including natural gas liquids and biofuels have allowed "all liquids" to increase slowly.

N.M. Ahmed, *Failing States, Collapsing Systems*, SpringerBriefs in Energy,
DOI 10.1007/978-3-319-47816-6_3

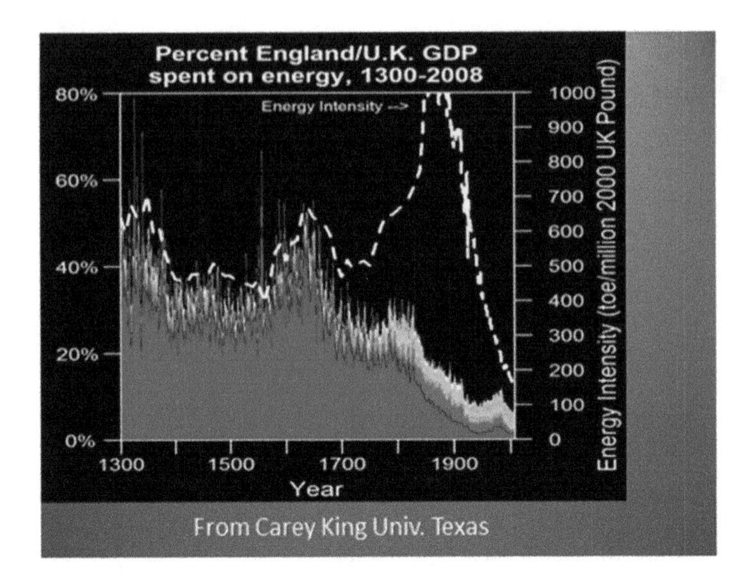

Fig. 3.1 Energy intensity in the UK *Source*: Carey King

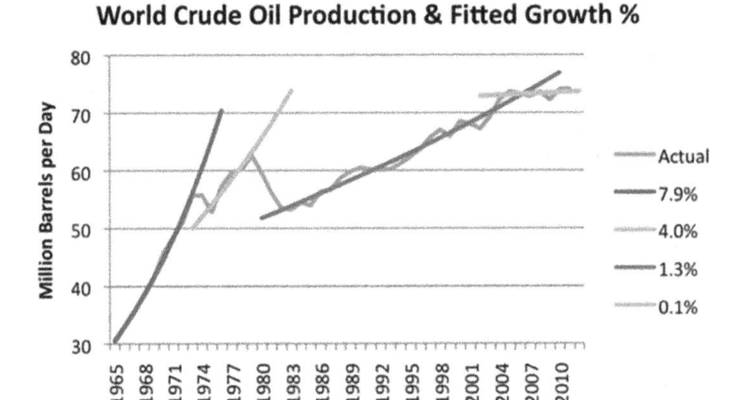

Fig. 3.2 Correlation between growth and oil production *Source*: Gail Tverberg

3.1 The Decline of Conventional Oil

Both oil industry leaders and the GMIC have largely obscured the economic implications of this by contesting traditional definitions of 'conventional' oil. By including forms of oil and gas which are technologically and economically more expensive to produce—such as 'tight oil' and natural gas—in total estimates of crude oil production, industry leaders like IHS Cambridge Energy Research Associates (CERA) argue that conventional oil production continues to increase, and will do so until 2030 and perhaps through to 2040. Despite this, they concede that:

> … there are severe long-term constraints beyond that… In the short term, it should be possible to continue capacity growth at levels necessary to meet the current pace of demand growth; but in the long term, the growth path will flatten and eventually turn downwards.
>
> With continuing strong long-term global demand growth (say 1% per annum), supply will soon become persistently tight and prices will respond accordingly. Resource scarcity, oilfield depletion, increasing upstream costs and persistently high oil prices will start to drive oil slowly out of the global energy equation. In a short time frame, any strong but volatile demand growth would not be met by the necessarily low levels of supply growth envisaged in this model. Because oil is ultimately a finite resource, supply growth by definition cannot be infinite – and we have seen that play out already, with long-term prices (US$80–100) that were once beyond imagination now considered to be reasonable. In practice, we can access portions of the ultimate resource base at ever higher costs, but at some point – well before exhausting the entire resource base – it will become more economical to use gas and alternative sources of energy, for example solar power. Over the past 10 years, we have entered a new, far higher cost base in the evolution of supply, which is a step in that direction (Jackson and Smith 2014).

However, there are considerable problems with this approach. The most obvious is that labelling a significant quantity of new liquids production under the rubric of 'conventional oil' despite fundamental differences in its energy quality is a misleading sleight-of-hand exercise that obscures more than it reveals. Tight oil, for instance, is defined as "compositionally" the same as crude oil, yet it requires expensive unconventional technologies to be applied to unconventional sources such as oil source rocks and other "hard-to-produce reservoirs."

Due to this shift to such sources, whether or not the resulting growth in production is defined as 'conventional', or encapsulated under an even wider estimate of overall 'liquids' production, what this sort of language obscures is the reality that since 2005, the increase in global liquids production has come from "hard-to-produce" sources, using expensive unconventional production techniques.

For this reason, the language of 'peak oil' alone, and ensuing GMIC claims that 'peak oil' forecasts are simply wrong because the world is on course to increase all-liquids production for several decades, has been unhelpful in assessing the thermodynamic reality of the actual net value of global energy production from hydrocarbon sources.

From an EROI perspective, the definitional categories used to delineate and describe various types of liquids production is irrelevant compared to the key question of the shift in the net value of energy produced from these sources. An EROI analysis, however, clearly demarcates the cheap and high quality crude oil production up to around 2004–2009, which has roughly plateaued (allowing for some very marginal increase), and the significant increase in more expensive, lower quality oil and gas production that has allowed all-liquids production to rise to meet growing world energy demand despite that plateau (Guilford et al. 2011).

Either way, whether such production is labelled 'conventional' or otherwise, the critical issue is that a particular type of higher quality, cheaper energy whose production has been increasingly constrained since 2005, has paved the way for a transition to geophysically more problematic forms of oil and gas which have a much lower energy value. That is why even the IHS CERA study admits that: "All categories of oil resource are now more expensive to develop, requiring high oil prices to generate an economic return" (Jackson and Smith 2014).

Whereas the EROI value of conventional oil is estimated at between 17:1 and 18:1, the EROI values of unconventional sources, like tar sands, shale oil, and shale gas is far lower: "The average value for EROI of tar sands is four. Only ten percent of that amount is economically profitable with current technology." For shale oil and gas: "The EROI varies between 1.5 and 4, with an average value of 2.8. Shale oil is very similar to the tar sands; being both oil sources of very low quality. The shale gas revolution did not start because its exploitation was a very good idea; but because the most attractive economic opportunities were previously exploited and exhausted" (Castillo-Mussot et al. 2016). As we will see below, shale gas's EROI is likely not this low, but still lower in practice than some believe.

3.2 The Rise and Decline of Unconventional Oil and Gas

In fact, some studies come to even starker conclusions. One commissioned by the investor coalition Ceres warns that production costs, market instability, and low EROI of less than a third of conventional oil's EROI, are endangering the viability of investments in unconventional oil (RiskMetrics Group 2010).

The Ceres study corroborates a Boston University analysis of the EROI of unconventional oil, finding it to be "extremely low" at between 1:1 and 2:1 when internal energy consumed in the oil shale conversion process is counted as a cost. An EROI of 1:1 means there is no energy "profit" from the investment of energy at all (Cleveland and O'Connor 2011).

In effect, the growing reliance on unconventional oil and gas has meant that, overall, the costs and inputs into energy production are rising inexorably. A comprehensive assessment of EROI of different energy sources commissioned by the UK government's Department for International Development (DFID) concluded, contrary to conventional industry wisdom, that ongoing decline in EROI values of fossil fuels is direct evidence that technological improvements have been unable to prevent accelerating depletion: "In conclusion, the EROI for the world's most important fuels, oil and gas, has declined over the past one to two decades for all nations examined." Moreover, the inclusion of different sources in assessments of the production and quality of 'conventional' oil has obscured the extent of the EROI decline: "It remains possible that the relatively high EROI values for the natural gas extracted during, and often used for, the production of oil may mask a much steeper decline in the EROI of oil alone." The DFID study thus forecasts that: "… it is probable that the EROI of oil and gas will continue to decline over the coming decades. The continued pattern of declining EROI diminishes the importance of arguments and reports that the world has substantially more oil remaining to be explored, drilled and pumped" (Hall et al. 2014).

Another sleight-of-hand offered by the IHS CERA study is the claim of continuously increasing global fossil fuel reserve sizes, due to the addition of new discoveries and improvements in technologies. Several studies, however, acknowledge that dramatic increases in estimates of global oil reserves are unlikely to be justifiable, and have little sound scientific basis.

In its *World Energy Outlook 2009*, the International Energy Agency (IEA) effectively conceded that the apparent doubling of world reserves since 1980 was politically-motivated, coming largely from upward revisions by OPEC countries "driven by negotiations at that time over production quotas and have little to do with the discovery of new reserves or physical appraisal work on discovered fields." (IEA 2009).

Two recent scientific reviews have corroborated this conclusion. One by the UK government's former chief scientific advisor, Sir David King, concluded that official estimates of world total oil reserves (including conventional, deepwater and unconventional resources) should be downgraded from 1150 to 1350 barrels to between 850 and 900 billion barrels (Owen et al. 2010). The other was authored by Michael Jefferson, former chief economist at Royal Dutch Shell Group, who reports that "the five major Middle East oil exporters altered the basis of their definition of 'proved' conventional oil reserves from a 90 percent probability down to a 50 percent probability from 1984. The result has been an apparent (but not real) increase in their 'proved' conventional oil reserves of some 435 billion barrels."

Global reserves have been further inflated, he concluded, by adding reserve figures from Venezuelan heavy oil and Canadian tar sands—despite the fact that they are "more difficult and costly to extract" and generally of "poorer quality" than conventional oil. This has unjustifiably inflated estimates of total global reserves by a further 440 billion barrels.

Jefferson's conclusion is stark: "Put bluntly, the standard claim that the world has proved conventional oil reserves of nearly 1.7 trillion barrels is overstated by about 875 billion barrels. Thus, despite the fall in crude oil prices from a new peak in June, 2014, after that of July, 2008, the 'peak oil' issue remains with us" (Jefferson 2016).

Jefferson and King provide compelling evidence directly undermining the IHS CERA thesis of oil abundance, that has in turn justified pushing forward expectations of the peak of global oil production to mid-century or after. Their analyses suggest that forecasts anticipating peaks in production emerging well before mid-century are far more accurate than the conventional industry scenarios that push forward the peak dates to after mid-century. This also establishes a geophysical basis for understanding the inexorable decline of EROI of traditional mineral energy sources, as being fundamentally related to the depletion of cheaper, higher quality sources far faster than conventional industry projections anticipate.

Several studies show, contrary to the more conventional optimistic projections, that unconventional oil and gas will be unable, due to this fundamental EROI constraint, to make up for the shortfall in conventional oil production, let alone dramatically exceed it. One such study forecasted that "oil production in countries with rich unconventional resources, such as Canada and Venezuela, will be higher in production than Saudi Arabia and Iran from 2050 to 2060," leading to a shift in global oil market power away from the Middle East (Matsumoto and Voudouris 2014)—which sounds significant until one factors in that a core reason the former production is higher is because production in Saudi Arabia and Iran at this point would be considerably lower.

In any case, this forecast derives from the ACEGES model, one of whose original contributors is Michael Jefferson. Ironically, the principal problem with the application of the model in the preceding study is that it completely ignores the caveat noted by one of its own contributors, Jefferson, to the effect that reserve sizes for unconventional oil and gas are heavily over-inflated.

Notably, the model's optimistic projections for unconventionals ignores the findings of earlier studies, such as by the Hydrocarbon Depletion Study Group at Uppsala University in Sweden. The group investigated the viability of a crash programme for the Canadian tar sands industry within the time frame 2006–2018 and 2006–2050, finding that even adopting "a very optimistic scenario Canada's oil sands will not prevent Peak Oil" (Söderbergh et al. 2007). Meanwhile as of 2016 as prices have declined from their peak the production of both Canadian and Venezuelan tar sands has dropped off considerably (Morgan 2016; Vyas and Puko 2016).

Similarly, a University of Newcastle study modelled a long-term prediction for unconventional oil production, and found that while unconventional oil would take a considerable amount of time to actually peak (between 2076 and 2084), rising unconventional production would be useless in mitigating the peak of conventional production: "If conventional oil production is at peak production then projected unconventional oil production cannot mitigate peaking of conventional oil alone" (Mohr and Evans 2010).

3.3 The Rise and Decline of Shale Gas

Similar concerns apply to shale gas. An extensive analysis by former Amoco petroleum geologist Arthur Berman, who has consulted for ExxonMobil and Total, challenges industry forecasts for shale gas. He argued presciently that actual shale gas production rates would be less than half of official industry projections—this is because production decline rates at shale wells are far higher than assumed. Although EROI of shale gas at the well head is high, the EROI of all gas production rapidly declines as energy costs of compression and distribution to consumers is factored in (Klump and Polson 2016). Industry officials frequently point to studies calculating EROI of "typical" wells, resulting in findings in the range of 64:1 and 112:1 (Aucott and Melillo 2013). The need for caution is obvious given that this is multiple factors higher than even conventional oil. In reality, studies focusing on EROI at the well head fail to offer a meaningful accounting of the further energy inputs required to transfer gas production at the well into usable electricity for the market. When such a wider analysis is conducted, the EROI of shale gas with respect to electricity generation for the consumer is much lower, at 12:1 or lower according to one study (Yaritani and Matsushima 2014), and at 10:1 according to a newer analysis putting it on parity with solar photovoltaics (Moeller and Murphy 2016).

This coheres with similar projections by David Hughes, former senior geoscientist for the Geological Survey of Canada, who incorporates consideration of technological developments for exploitation of unconventional gas and concludes that world gas production will nevertheless peak around 2027 (Homer-Dixon 2011). A separate study similarly found that although the interplay between technology and prices for unconventional gas could significantly widen the possible range in which peak gas production may occur to from 2019 to 2062, with the best estimate of peak production is around 2028: "While it was found that the production of unconventional gas was considerable, it was unable to mitigate conventional gas peaking" (Mohr 2010; Mohr et al. 2015).

3.4 The Decline of Coal and Uranium

Likewise for coal. While coal reserves exist in abundance worldwide, the depletion of high quality resources that are cheap to exploit (i.e. high energy content and with little overburden and in thick seams) has meant that remaining reserves are more expensive to extract and lower quality in terms of the net energy they produce. A major country-by-country analysis projects a high probability of world coal production peaking "on an energy basis" in 2026. The overall range of probability means the peak could occur later, by 2047 (Mohr and Evans 2009).

With hydrocarbon energy resources facing inexorable depletion over coming decades—no matter where one stands on the projections as an optimist or pessimist—nuclear energy often seems an attractive option. Yet the problem of net energy decline applies here, too.

One extensive study finds that the construction, mining, milling, transporting, refining, enrichment, waste reprocessing/disposal, fabrication, operation and decommissioning processes of nuclear power are heavily dependent on fossil fuels (Pearce 2008). This raises serious questions about the viability of nuclear power in about two decades time, when hydrocarbon resources are likely to be well past their production peaks. Further, the study concludes that nuclear power is simply not efficient enough to replace fossil fuels, an endeavour which would require nuclear production to increase by 10.5 % every year from 2010 to 2050—an "unsustainable prospect". This large growth rate requires a "cannibalistic effect", whereby nuclear energy itself must be used to supply the energy to construct future nuclear power plants. The upshot is that the books cannot be balanced as the tremendous amounts of energy necessary for mining and processing uranium ore, building and operating the power plant, and so on, cannot be offset by output in a high growth scenario. In particular, growth limits are set by the grade of uranium ore available—and high-grade uranium is predicted to become rapidly depleted in coming decades, leaving largely low-grade ore falling below 0.02 % (Pearce 2008).

3.5 Peak and Terminal Decline of Net Energy

It should be noted, however, that debate over the precise point of peak production of an energy resource can be ultimately misleading by itself. Production levels in themselves do not reflect the quality of the energy being produced as measured by

EROI. Therefore, an analysis focusing purely on rates and levels of production can in fact mask the underlying dynamics of accelerating production costs, driven by a decline in resource quality, which means that rising production is unable to meet the energy requirements of society due to plummeting EROI values.

This is shown in a major Royal Society study focusing on the declining EROI for global oil and gas production, which calculated that the latter's EROI is roughly 15:1 and declining. Looking closer, the study found that for the US, EROI of oil and gas production is about 11:1 and declining; and for most unconventional oil and biofuels is less than 10:1. It concludes: "… as the EROI of the average barrel of oil declines, long-term economic growth will become harder to achieve and come at an increasingly higher financial, energetic and environmental cost" (Murphy 2014).

What is particularly important to note here is that since the 1960s, EROI has declined, even as production has continued to increase, though its rate of increase has also declined and appears to be approaching plateaus on a number of fronts. This highlights the often overlooked relationship between EROI and production that underscores the extent to which seemingly accelerating production in the short-term, can itself be symptomatic of a geophysically catabolic process in the long-term.

As EROI declines, increasing costs are required to maintain production. As demand rises due to economic growth, production must increase to meet demand, therefore requiring further financial, energetic and environmental inputs. But to increase production to a level sufficient to meet demand while EROI is declining, means that even more quantities of energy are required simply to maintain production, and even higher quantities required to increase it. Accelerating inputs equate to an escalating cost that, in turn, culminates in a further decrease in EROI even while driving increased production. The trajectory of increasing production conventionally hailed by the industry as *prima facie* evidence of its barometer of health is in fact the prime indicator of the catabolic depletion of the resource in question (Murphy and Hall 2011).

This is why it is perhaps more useful to speak of the peak and decline of EROI as a measure of the health of the global energy system, and indication of its future trajectory. According to this measure, EROI for global oil and gas production, for instance, peaked in 1999 and has since entered terminal decline (Gagnon et al. 2009). Notably, in the same period, while oil and gas production has increased, the rate of increase has dramatically slowed compared to previous decades. Simultaneously, capital expenditures (capex) by the world's oil majors has over the same period increased by over 10 % every year, directly illustrating the self-catabolic process at play, consisting of heightening efforts to exploit the resource base resulting in declining EROI and flattening production (Brandt et al. 2015).

The self-catabolic nature of this process can be explored from multiple interconnected angles. The first we will discuss below concerns the direct impact of such escalating costs on the global economy, manifesting in an emerging limit to growth. The second is the direct environmental cost of the increasingly rapid consumption and depletion of traditional mineral energy sources, which is having an increasing impacting on climate, oceans, and the global food system due to the thermodynamic consequences in uncontrolled energy releases. The third concerns the amplifying feedbacks between these indefinite antecedent domains of crisis, resulting in an unmistakable deterioration in state power and an acceleration of political violence over the same period.

Chapter 4
Permanent Secular Stagnation

4.1 Empirical Blindspots of Economic Theory

Over the last few centuries, the 'progressive' trajectory of human civilization can be encapsulated in data for energy consumption and GDP.

As would be expected from the preceding discussion, an issue largely overlooked within conventional economic theory is that all economic systems are fundamentally physical systems in which energy is transmitted and converted into different forms, in compliance with the second law of thermodynamics—and in which depletion must take place with respect to non-renewable energy resources due to the conservation of matter and energy.

Economic theory, however, does not recognise that the laws of physics provide any meaningful constraint on the capacity of economies to grow continuously by forever increasing their material throughput. While there is scope to recognise that economic systems manifest "emergent properties" with their own distinctive rules, patterns and structures, that does not mean that these rules can be trivially deduced from—or break—the laws of physics. Rather, as with any complex system, the macro-structures emerge from but still operate within those laws. This means that however "emergent" the macro-structures of economics appear to be, they can never be free of the second law of thermodynamics (Pueyo 2014).

Thus, the direct correlation between economic growth and the growth of energy consumption is because the former is fundamentally dependent on and enabled by the latter.

4.2 Economy as Embedded in Energy

Human population, the global economy, and energy production have all "grown exponentially", finds a major study for the American Institute of Biological Sciences—yet energy still "imposes fundamental constraints on economic growth

N.M. Ahmed, *Failing States, Collapsing Systems*, SpringerBriefs in Energy,
DOI 10.1007/978-3-319-47816-6_4

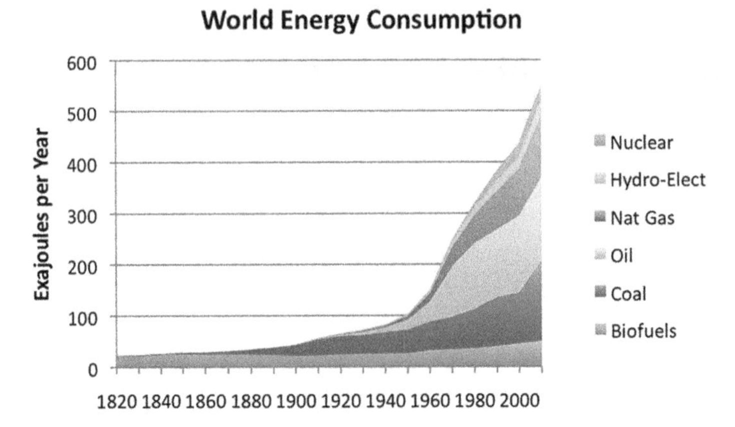

Fig. 4.1 Increasing Global Energy Consumption. *Source*: Gail Tverberg

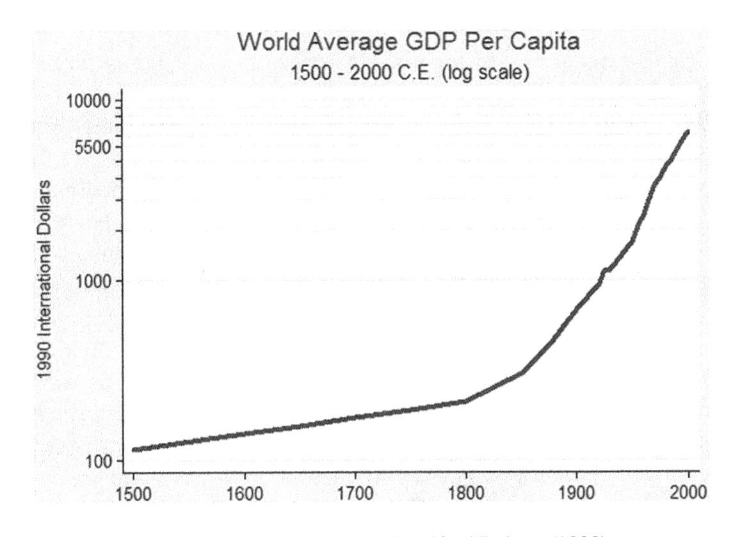

Fig. 4.2 Exponentially increasing GDP. *Source*: J. Bradford Delong (1998)

and development." The study integrates data across physics, ecology and economics to "demonstrate a positive scaling relationship between per capita energy use and per capita gross domestic product (GDP) both across nations and within nations over time. Other indicators of socioeconomic status and ecological impact are correlated with energy use and GDP" (J. H. Brown et al. 2011) (Figs. 4.1 and 4.2).

The two graphs above illustrate the direct relationship between economic growth and energy production during the period of the emergence and expansion of industrial civilization up to around 2000.

A further graph below illustrates the direct correlation between GDP and energy consumption for the period between 1969 and 2013 (Fig. 4.3):

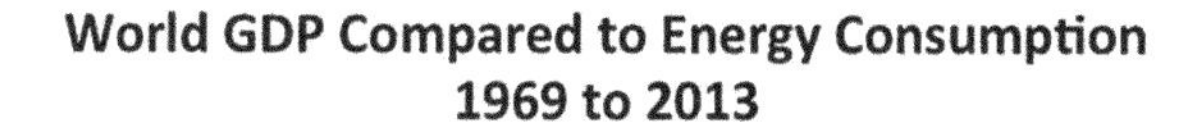

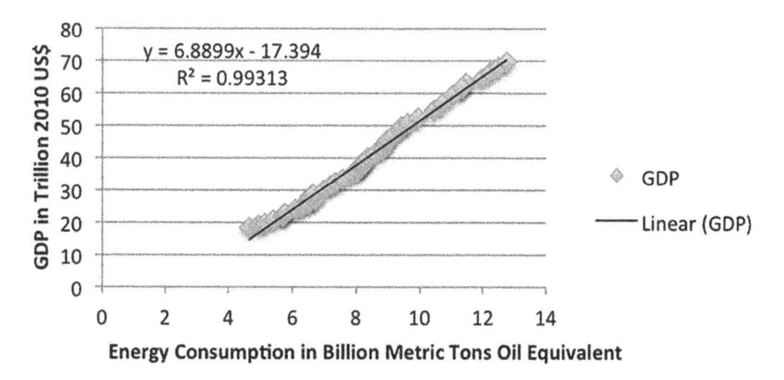

Fig. 4.3 Correlation between Global Energy Consumption and GDP. *Source*: Gail Tverberg

Despite the empirical evidence for their conclusions, though, the authors of the *Bioscience* paper lament that "these perspectives have not been incorporated into mainstream economic theory, practice, or pedagogy and they have been downplayed in consensus statements by influential ecologists and sustainability scientists." (J. H. Brown et al. 2011).

4.3 Plateau and Decline of Economic Growth

As a result, conventional economists failed to anticipate the 2008 global financial crisis, and since then have consistently failed to anticipate the major economic crises of ensuing years, while consistently and incorrectly forecasting returns to economic growth. In reality, however, economic growth on a global scale is experiencing an unmistakeable plateau, that correlates clearly with the emerging plateau in energy production (Fig. 4.4).

According to Jancovici, since the 1960s—which is when the EROI of the global fossil fuel system as a whole was at its highest according to most studies: "...the growth rate of the GDP per capita (world average) has been slowly—and constantly—decreasing..." (Jancovici 2013). In the decade after 1960, he calculates, GDP was increasing at +3.5 % per year. For the decade after 1970, this rate of increase dropped to +2 % per year. Over the last three decades, the rate of GDP growth dropped to +1.5 % per year. And in the period following the 2007–8 financial crash up to 2012, it has dropped even further to 0.4 % per year on average (Fig. 4.5).

The steady decline in the rate of GDP growth thus correlates directly with the steady decline in EROI of production from the global fossil fuel resource base, even as energy production has continued to increase. However, as energy production has slowed down over the last decade since 2005—accompanying the shift to lower quality unconventional liquids—now approaching an undulating plateau, so too has GDP growth.

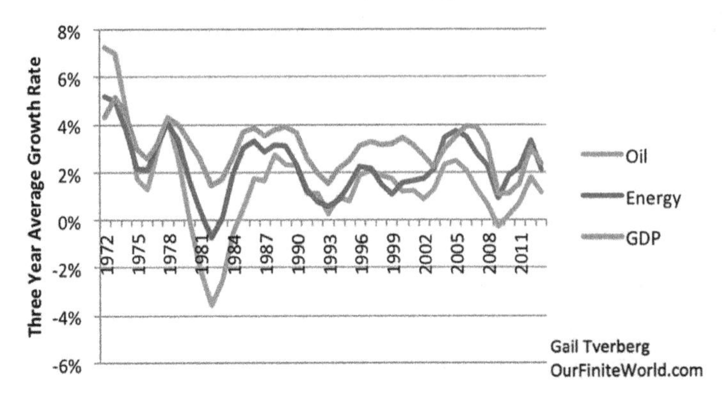

Fig. 4.4 Correlation between oil, energy and GDP. *Source*: Gail Tverberg

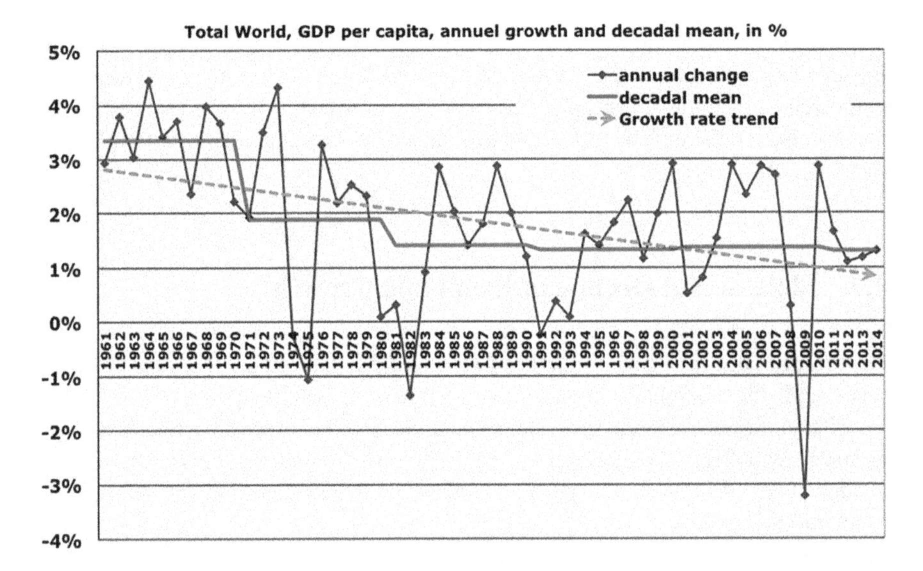

Fig. 4.5 Declining rate of economic growth. *Source*: Jean-Marc Jancovici

Consider this revealing diagram produced by Bloomberg, which illustrates how World Bank data confirms not only that economic growth is plateauing, but that it is likely to continue plateauing for the foreseeable future (Fig. 4.6):

4.4 The Mythology of Decoupling

Despite the verdict of BP to the effect that a significant decoupling between energy production and economic growth is already taking place due to increasing energy efficiency, finer-grained studies looking closely at the relevant data come to quite

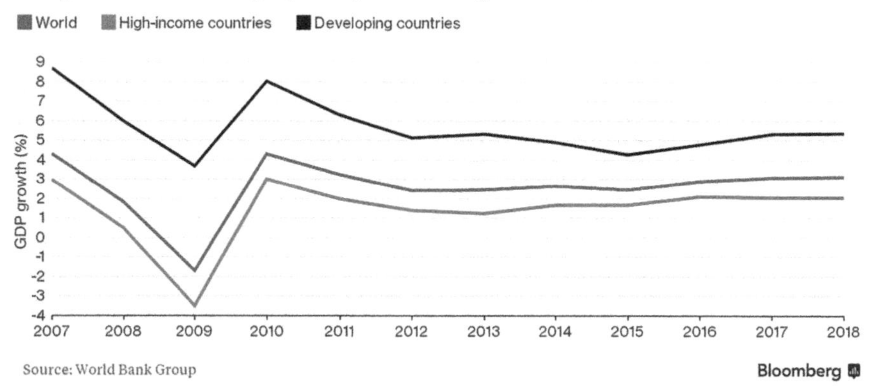

Fig. 4.6 Plateauing of economic growth. *Source*: Bloomberg

different conclusions. BP has argued that energy intensity, the quantity of energy required per unit of GDP, will decline by 36 % from 2012 to 2035 (Kaminska 2014).

Outside of the oil industry, the jury is still out on whether decoupling in this form can actually take place. An independent econometric study in *Energy Economics* finds instead "that the growth of per capita energy use has been primarily driven by economic growth, convergence in energy intensity, and weak decoupling. There is no sign of strong decoupling" (Csereklyei and Stern 2015). The most compelling critique of the decoupling mythology was published in the *Proceedings for the National Academy of Sciences*, which found that data used to claim success in decoupling ignored the role of resource consumption in the expanding role of international trade. Using a consumption-based indicator of resource use known as a Material Footprint (MF) framework, the study found that: "Achievements in decoupling in advanced economies are smaller than reported or even nonexistent… countries' use of nondomestic resources is, on average, about threefold larger than the physical quantity of traded goods. As wealth grows, countries tend to reduce their domestic portion of materials extraction through international trade, whereas the overall mass of material consumption generally increases. With every 10 % increase in gross domestic product, the average national MF increases by 6 %" (Wiedmann et al. 2015).

Indeed, analyses like that of BP's ignore extensive evidence that economic growth, to the extent that it has been able to continue, is being driven largely by an increasing availability of cheap credit—rather than any fundamental and permanent transformations in energy intensity. The ratio of global debt, excluding financial institutions, has grown from 175 % of global GDP on the eve of the 2007/2008 financial crisis to 210 % today. Cheap credit has enabled excessive borrowing, risk taking and sharply rising asset prices, driving the same form of unsustainable debt-driven growth that partly led to the 2008 financial crash (Stewart 2015).

For this reason, the seeming decoupling between energy production and GDP growth detected by analysts at BP and elsewhere is illusory, and conceals the extent to which growth, especially since the 1970s, has been premised increasingly on the financialization of the economy through the creation of new instruments of credit creation to permit extensive leveraging. Such debt-driven growth, however, only offsets the apparent biophysical limits to growth by accelerating debt and socializing the costs in the event of a financial crisis onto general consumers, while protecting the financial institutions most responsible for debt-generation. In the energy sector, as oil prices have slumped, growth has increasingly been driven by debt. Oil majors ExxonMobil, Royal Dutch Shell, BP, and Chevron hold a combined net debt of $184 billion, more than double their 2014 debt levels (Williams and Olson 2016).

Global industrial civilization is thus facing a convergence of crises: the plateauing of energy production and the plateauing of economic growth, amidst an inexorable decline in EROI. What is the long-term prospect for this convergence? Two recent studies have attempted to model this convergence and forecast the future of economic growth in the context of energy production.

The first by a team at the Institute for Marine Sciences in Barcelona forecasts a peak in total fossil fuel production in 2038, which "may cause shortages in energy supplies and major disturbances in the global economy." The team argues that the most plausible scenario is that economic growth will eventually permanently plateau after the fossil fuel production peak, leading to a "stationary economy"—even while populations and aspirations increase (García-Olivares and Ballabrera-Poy 2015).

Another study by an interdisciplinary team of Spanish scientists and economists came to similar conclusions (Capellán-Pérez et al. 2014). Using an Economy-Energy-Environment model based on System Dynamics, the team attempted to integrate an overall forecast of net energy production from oil, gas, coal and uranium: "The results show that demand-driven evolution, as performed in the past, might be unfeasible: strong energy-supply scarcity is found in the next two decades, especially in the transportation sector before 2020. Electricity generation is unable to fulfill its demand in 2025–2040, and a large expansion of electric renewable energies move us close to their limits." The team concluded that the most plausible economic scenarios associated with these outcomes in the latter half of the century are "zero or negative economic growth" (Capellán-Pérez et al. 2014).

Ironically, then, the very exponential acceleration of growth over the last few centuries—peaking over the last few decades—has masked the core catabolic geophysical processes that have enabled this process in the short-term, while simultaneously disabling its indefinite continuation in the long-term.

Yet the immediate self-catabolic economic costs of ongoing economic growth encompass only one dimension of the overarching costs of growth. The other pertains to the wider environmental systems in which both the global energy and economic systems are embedded.

Chapter 5
Earth System Disruption

It is sometimes assumed that due to the trajectory of fossil fuel depletion, the world will be able to avoid dangerous climate change simply because we will 'run out' of oil before carbon dioxide emissions reach a tipping point. This, however, is based on an incomplete understanding of the vulnerability of the Earth System to climate disruption.

The dark side of economic progress is perhaps most amply illustrated in the direct correlation between exponential economic growth and the growth of greenhouse gas emissions. The over-dependence (by which is implied a sense of structural dependence that cannot be overcome) of economic growth on fossil fuels has meant that the more fossil fuels are burned by human civilization along this path of accelerating growth, the greater the quantity of carbon dioxide and other greenhouse gas emissions are inputted into the atmosphere, leading to the gradual unbalancing of the natural carbon cycle (Fig. 5.1).

As the Earth System is itself a complex system, minor perturbations in the system can have wide-ranging impacts that are not easy to anticipate or forecast. Climate models are forever having to play catch-up due to the inherent complexity of the Earth's ecosystems and the difficulty in capturing all the relevant variables.

In the case of climate change, we are now beginning to grasp that climate change is perhaps the largest and most potent overarching impact of the current trajectory of human civilization on the Earth System. The more we understand the tightly coupled nature of these various sub-systems in the Earth System, the more we are beginning to grasp how dangerous the impacts could be.

Rockstrom et al. first mobilized the concept of "planetary boundaries" to delineate what appeared to be a "safe operating space" within the Earth System necessary for humanity to survive. Those boundaries, they found, are at risk of being breached due to anthropogenic climate change, which could in turn trigger abrupt and potentially catastrophic environmental changes (Rockström et al. 2009). An update of this landmark research concluded that four of the nine planetary boundaries identified had already been breached: climate change, species loss, land-use

N.M. Ahmed, *Failing States, Collapsing Systems*, SpringerBriefs in Energy,
DOI 10.1007/978-3-319-47816-6_5

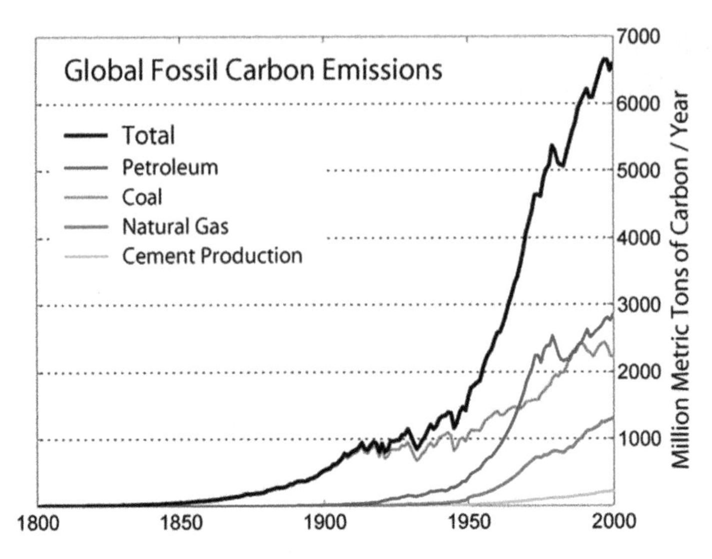

Fig. 5.1 Global carbon emissions from fossil fuels. *Source*: Wikimedia

change, and altered bio-geochemical cycles from overuse of fertilizers (Steffen et al. 2015).

To illustrate the complexity of these issues, we will focus here on three phenomena directly related to accelerating anthropogenic climate change, demonstrating the extent to which human civilization is already experiencing the impacts of increasing Earth System disruption: ocean acidification; heatwaves; and extreme weather events.

5.1 Ocean Acification

As climate change is accelerating, so is the acidification of the oceans. The two processes are causally related. Oceans are becoming more acidic because the vast bulk of global warming due to climate change is absorbed into the oceans, both in terms of the increase in temperatures, and in terms of actual carbon dioxide emissions. The massive increase in CO_2 levels in the oceans means that ocean pH levels are dropping dramatically. In a high emissions scenario—which characterizes our current business-as-usual trajectory—scientists cannot rule out further mass extinctions of marine life (Azevedo et al. 2015) (Fig. 5.2).

If this trend continues at current rates, before the end of the century, the scale of ocean acidification will threaten vast interconnected webs of marine life on which millions of people are dependent for their food supply and livelihoods. A major study in this regard notes that fully 80 % of animal protein consumed in the world comes from fish. Yet at current rates, by 2100, an estimated 98 % of the oceans will be affected by a combination of deleterious climate impacts including acidification,

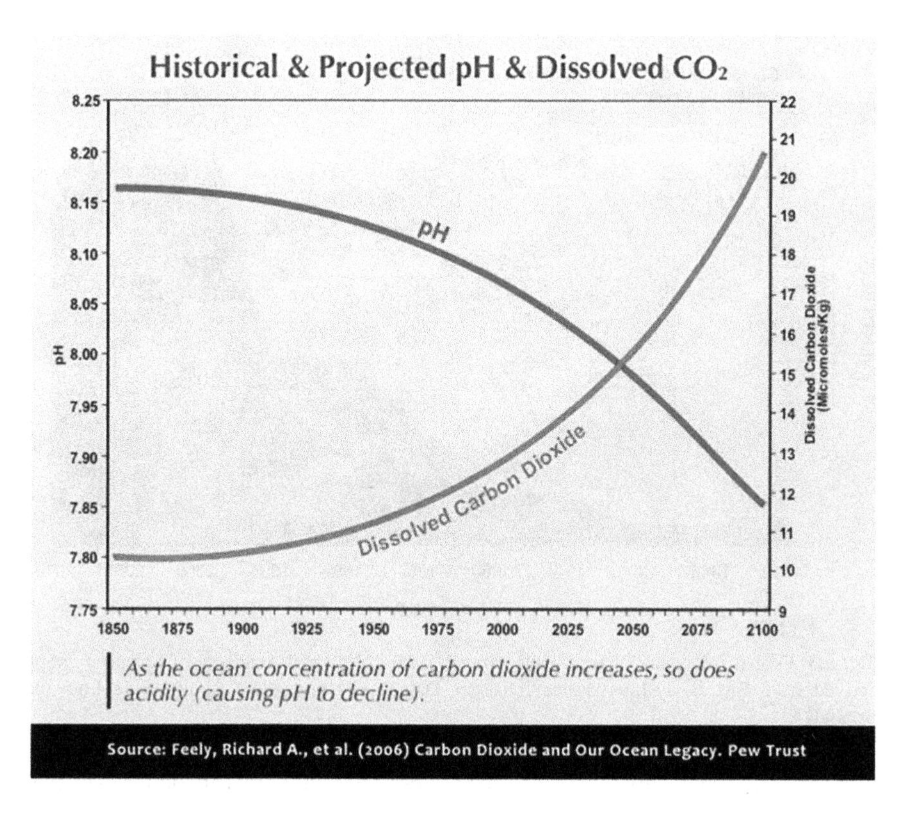

Fig. 5.2 Historical and projected ocean acidification. *Source*: Richard Feely (2006)

low oxygen, and high temperatures. "These results highlight the high risk of degradation of marine ecosystems and associated human hardship expected in a future following current trends in anthropogenic greenhouse gas emissions," conclude the study authors (Mora et al. 2013b).

Ocean acidification has also been discovered to be a prime driver of the largest mass extinction event in the history of the planet, the Permian-Triassic extinction event. The rate at which carbon was released during that event is similar to the rate of modern carbon emissions (Clarkson et al. 2015).

5.2 Heat Waves

While the oceans are dying, above the oceans the atmosphere is already experiencing the direct impact of climate change in the form of intensifying heatwaves and extreme weather events. The increasing frequency—and increasing intensity—of heat waves is perhaps one of the most overt manifestations of the dangerous impacts of climate change. Since 1950, the number of heat waves worldwide has increased,

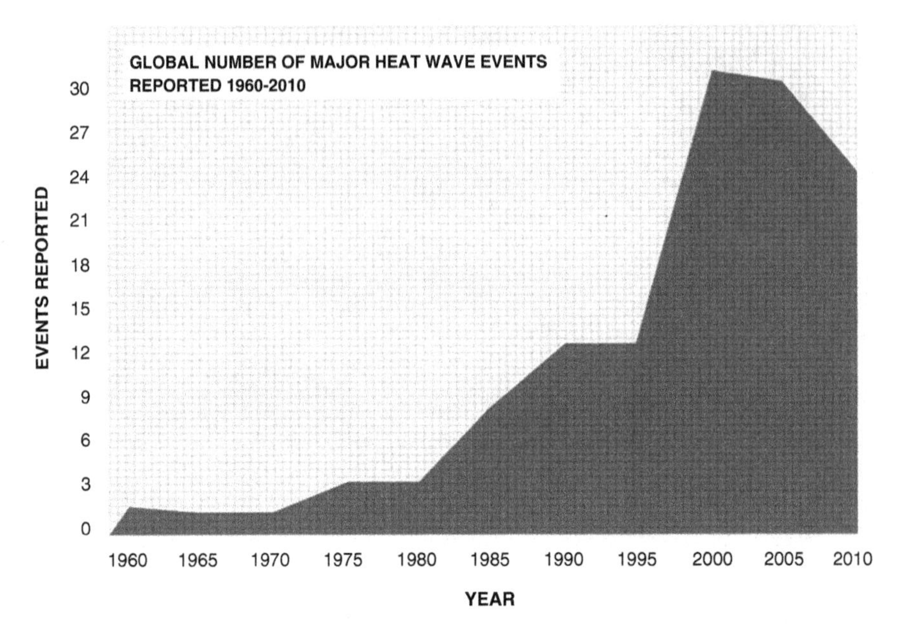

Fig. 5.3 Global frequency of heat wave events. *Source*: National Academy of Sciences, graph derived from EM DAT, International Disaster Database, Universite Catholique de Louvain, Brussels

heat waves have become longer, and the hottest days and nights are hotter than ever before. In recent years, the global area affected by summer heatwaves has increased 50-fold. Within the US, the direct impact of more frequent and intense heatwaves is an increasing frequency and duration in wildfires (Trendberth et al. 2012) (Fig. 5.3).

The increase in heat waves has accompanied a similar increase in the frequency and intensity of extreme weather events more generally, and relatedly, of natural disasters many of which are directly related to extreme weather. The last half century has seen a dramatic increase in the frequency and severity of extreme weather events in the form of droughts, wildfires, extreme rainfall, floods, hurricanes and tornadoes. The Met Office concludes that despite scientists' reluctance to attribute specific extreme weather events to human-induced climate change, there is now no longer any doubt that climate change is making extreme weather increasingly likely all over the world (Stott 2016) (Fig. 5.4).

By far the most disturbing study led by the University of Hawaii argued that the pattern of escalating intensity and frequency indicates that anthropogenic climate change is rapidly pushing the climate system into a 'new normal', that breaks fundamentally with the preceding 150 years. The paper came up with the concept of "climate departure" to explain its prediction that in coming decades, the trajectory of escalating extreme weather signals that the climate is destined to 'depart' from the historical norm of weather as we have known it. On a business-as-usual trajectory, the initial locus of this "climate departure" will occur *within the next decade* in

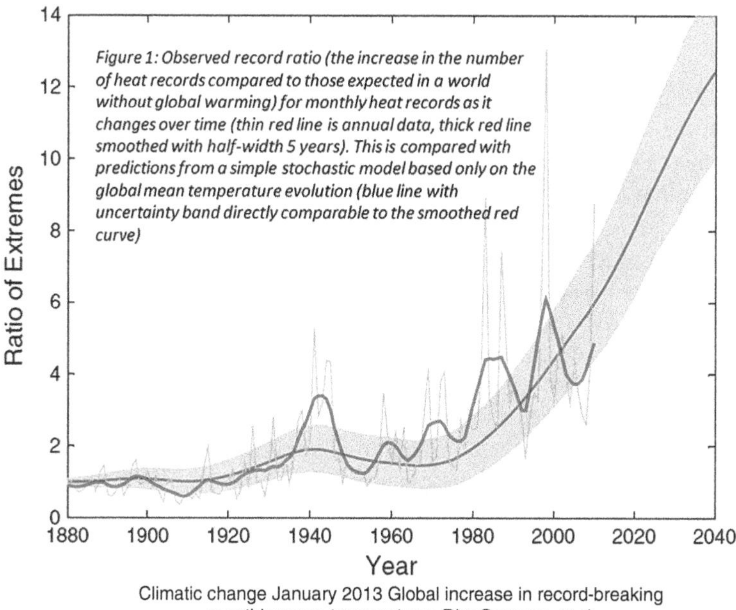

Fig. 5.4 Global increase in record-breaking monthly average temperatures. *Source*: Dim Coumou et al. (2013)

the tropics—that is, a vast region encompassing parts of the Middle East, Central Asia, South Asia and Africa. On a global scale, "climate departure"—the entry into a 'new normal' of extreme weather—will hit around 2047. Even under stringent carbon emission mitigation scenarios, this tendency to "climate departure" will not be halted—only postponed a few more decades, to around 2069 (Mora et al. 2013a).

This dire verdict was corroborated in a later study led by the Max Planck Institute, which forecasts that even if climate change is stabilized at an average rise of 2 °C—the target adopted by international policymakers to avoid dangerous global warming—by 2050 (around the same date pinpointed by Mora et. al), the Middle East and North Africa would face prolonged heatwaves and dust-storms of such severity as to render much of these regions "uninhabitable". Temperatures during the summer in these regions will increase more than two times faster compared to the average global warming. South of the Mediterranean, temperatures would reach as high as 46 °C (approximately 114 °F) by mid-century, and five times more often than in the late 1990s. Heatwaves would likely occur ten times more than they do now. Such intolerable conditions would endanger the lives of the regions' 500 million inhabitants, and force people to migrate simply to survive (Lelieveld et al. 2016).

This means, very simply, that no matter what mitigation efforts look like on climate change, the coming decades will see increasing instability in the Middle East and North Africa, and an ever greater exodus from parts of the region into the Northern hemisphere. Intensifying climate-induced droughts and heatwaves will create conditions that no regional state will be able to cope with.

5.3 Food Production

But this trend is already dramatically affecting the global food system. Many of the extreme weather events in recent years have been concentrated in some of the world's most critical food basket regions, contributing directly to prolonged crop failures that have been linked to global food price spikes and other phenomena.

It is already known that anthropogenic climate change to date has had a debilitating impact on global food production, partly associated with the impact of more frequent extreme weather events on crop production. Total losses in national cereal production from 1964 to 2007 due to droughts and extreme heat likely caused or exacerbated by climate change have been estimated at 9–10 % (Lesk et al. 2016).

A new model created by British researchers—FEEDME (Food Estimation and Export for Diet and Malnutrition Evaluation)—has attempted to forecast the potential impacts of climate change on global food production, accounting for a range of different socio-economic and emission mitigation scenarios. The project was developed under the QUEST-GSI program led by the Walker Institute for Climate System Research at the University of Reading. Its astonishing forecast is that by mid-century, under a business-as-usual scenario, *more than half of the entire projected global population* will suffer from undernourishment due to a combination of climate change and inequality. The model explicitly does not account for technological or agricultural adaptations in order to explore future scenarios based on current policies. The specific breakdown is that without climate change, 31 % (2.5 billion people) of the global population will be "at risk of undernourishment" by 2050: "An additional 21 % (1.7 billion people) is at risk of undernourishment by 2050 when climate change is taken into account" (Dawson et al. 2014).

Multiple studies thus pinpoint mid-century as a major turning point in the global climate system, in which the climate enters an extreme 'new normal' radically distinctive from the environment that the human species, not to mention innumerable other species, have become accustomed to in recent centuries. The bulk of this 'new normal' will unfold and impact human societies regardless of climate mitigation efforts, and will consist of a new scale and order of atmospheric and oceanic disruption as global warming manifests in terms of extreme weather and ocean acidification. As we enter this new era of climate crisis, the costs to our economies will also accelerate. Compounded by the mounting costs of accelerating energy depletion, and the mounting costs of an escalating global food crisis, the economy will essentially face a decadal 'triple crunch' peaking in mid-century from a convergence of energy, climate and food crises.

Chapter 6
Human System Destabilization

The concept of a 'triple crunch' was first popularized by the New Economics Foundation (Simms 2008), focusing on high oil prices, the debt-induced credit crunch, and climate change—however, the analysis centered on the interconnected dynamic of what were ultimately *symptoms* of crisis, rather than identifying their core causal dynamics. The idea of a 'triple crunch' encompassing energy, climate and food crises must therefore be extended in the recognition, highlighted by the work of New Economics Foundation, of the role of credit—debt-money (the creation of money through the expansion of debt) as a key instrument used to sustain economic growth through the extension of the 'financial system.' While the abundance of cheap fossil fuels played the key role in permitting the expansion of the monetary and financial system—enabling exponential economic growth—from the 1950s onwards, the accelerating reduction in EROI has accompanied an increasing reliance on financialization: the shift from the expansion of money, to the expansion of credit (debt-money). Beginning concertedly in the 1970s, this has been most exemplified in the US Federal Reserve's post-2008 rubber-stamping of quantitative easing to use money printing or credit creation (debt-money expansionism) as a mechanism to offset economic crises and bailout insolvent banks endangered by mass consumer defaults. The policy's fiscal twin is austerity—clamping down on state expenditures in the form of public spending on infrastructure, education, health care and other forms of critical social investments and public services, while using state power to protect ongoing debt-based profiteering in the corporate-financial sectors (Smith-Nonini 2016).

The neoliberal era, with its policies of extreme deregulation, debt-money expansionism and harsh national austerity, is thus a direct product of the changing dynamics of the global energy system and the transition into a world of more expensive, lower quality, and environmentally more destructive fossil fuels.

As the 'triple crunch' tightens from now to mid-century, this forces the global economy as a whole into adopting a limited range of policies to address the dampening of growth underpinned by geophysical realities. Within a framework that continues to be wedded to the idea that business-as-usual must continue—one tied into

N.M. Ahmed, *Failing States, Collapsing Systems*, SpringerBriefs in Energy,
DOI 10.1007/978-3-319-47816-6_6

protecting the vested interests whose tremendous power remains embedded within the unequal structures of the prevailing geopolitical, economic and energy order—the only options are to rely excessively on debt-money expansionism to shore-up dying industries in the fossil fuel and financial sectors, while intensifying austerity so as to socialize the costs of this onto the wider public, while privatizing the 'benefits' in terms of profits. Those profits, however, will accrue to an ever tightening circle of financial and corporate institutions as more and more of the energy-financial incumbency are squeezed out under the weight of their own unsustainability. The escalation of national austerity policies cannot, however, be sustained for long without increasingly debilitating impacts on the health and well-being of wider publics. This means that phenomena such as the 2008 Occupy movement and the 2011 Arab Spring were not just historical blips that may or may not occur again, but represented major breaking points in the system due to populations feeling unable to adjust to intolerable conditions imposed by escalating global systemic crises. Such breaking points, then, represent a taste of things to come.

Supporting this conclusion is the fact that corresponding to the rising trends of increasing climate disruption and energy decline, recent decades have seen a marked increase in political violence worldwide. These outbreaks of political violence demonstrate that prevailing national state institutions, and their domestic monopolies in the means of violence (which is the basic underpinning of state power as defined by the capacity to mobilize violence to control a defined national territory) are increasingly being challenged and undermined. In other words, what we are witnessing is a creeping acceleration of the forces of non-state political violence that directly weaken the very fundamentals of state power.

We can track these forces using data on a range of different types of political violence: intra-state conflict; civil unrest; Islamist terrorism; and far-right terrorism.

Across the spectrum, the data shows that all these forms of violence have been intensifying especially since the 1970s, with a degree of fluctuation, with a particular acceleration since the late 1990s. This is significant, because 1999 is the year at which the global EROI of fossil fuels reached its total peak level, after which overall global EROI has increasingly and steadily declined.

We cannot draw reductive conclusions about this correlation. The causes of conflict or political violence cannot be ever reduced to one simplistic set of causes, and this is abundantly clear from the data itself. However, the data also indicates the need for a deeper explanation for a long-term upward trend in non-state political violence that goes beyond the surface of conventional political and geopolitical analysis.

6.1 Intrastate Violence

Data on the frequency of armed conflicts between 1946 and 2012 (Fig. 6.1) demonstrates an unrelenting trend of declining interstate conflict. Much of this has been addressed within the international relations literature with respect to factors such as

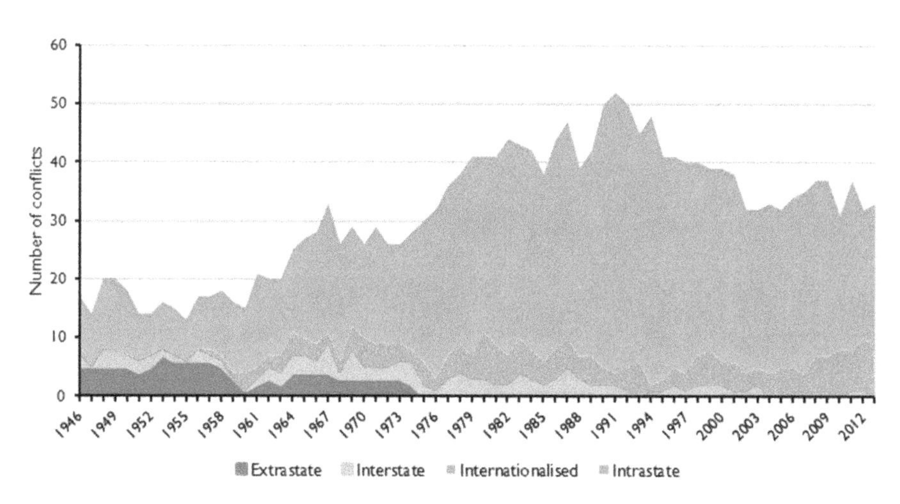

Fig. 6.1 Frequency of armed conflict. *Source*: PRIO Conflict Trends

the deterrent role of nuclear weapons and proliferation, the shift in political and popular culture in the context of the devastating experience of conventional wars, and the role of 'risk-transfer militarism' in which Western governments seek new technologies to transfer risks of conflict onto subject populations to avoid the massive fallout from putting soldiers at risk, leading to military interventions that rarely involve confrontations between militarily equal powers (Shaw 2005).

However, what is less easy to explain is the prevalence of intra-state conflict, that is, conflict within states. The data shows that intra-state conflict rose during the Cold War, which provides a plausible geopolitical context for this trend, then peaked in the early 1990s—shortly after the collapse of the Soviet Union—and since then declined. However, since 9/11, there appears to have been a fluctuating resurgence in the occurrence of intra-state conflict.

Greater clarity can be gleaned from data on the number of casualties from the different forms of conflict. Figure 6.2 illustrates an upward trend in casualties from intra-state conflict since 1946. The number of deaths from intra-state conflict as of the 1990s hit record levels, which have overall continued to increase.

6.2 Civil Unrest

Data on civil unrest also bears out these conclusions. A report by the Risk Advisory Group commissioned by Lloyd's insurers confirms that an analysis of "all types" of political violence over the period 1960–2013, demonstrates an overall upward trend (Fig. 6.3).

The study examines a wider range of data, looking at intra-state violence, as well as mass protests and civil unrest, finding that while there was a clear overall spike in the late 1980s to early 1990s—definitively related to Cold War geopolitics, the

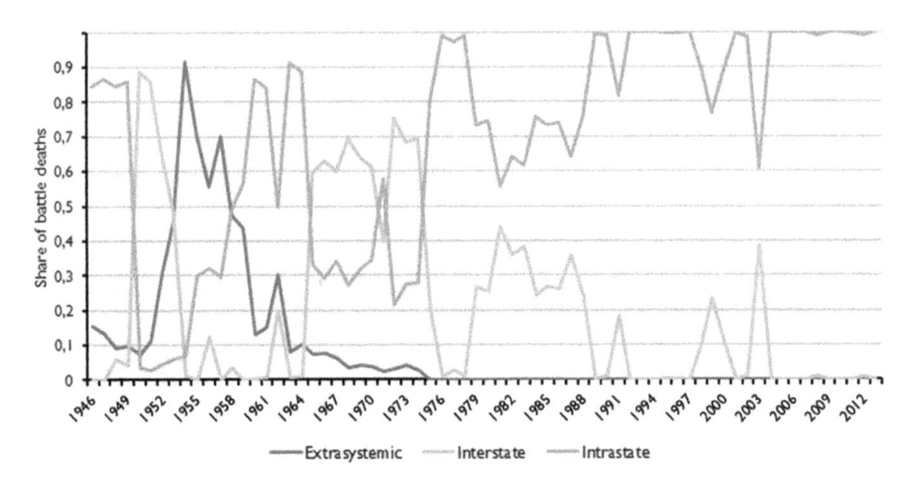

Fig. 6.2 Number of battle deaths as a proportion of total for different types of conflict. *Source*: PRIO Conflict Trends

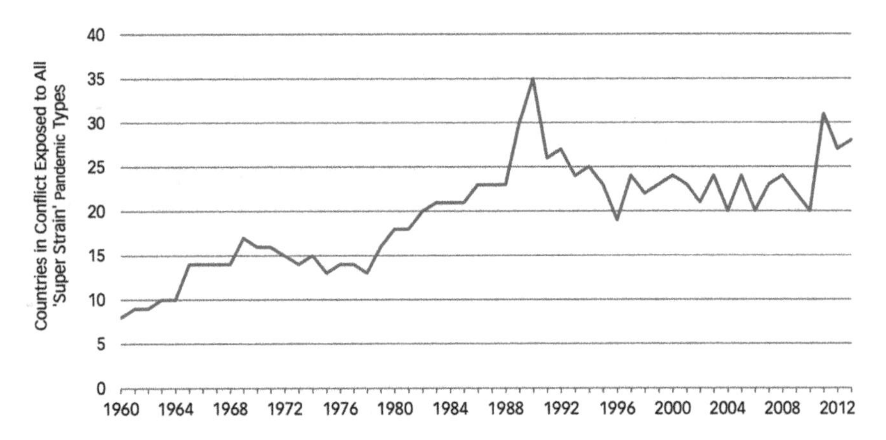

Fig. 6.3 Frequency in political violence pandemics

ensuing undulating plateau in political violence gave way after 2008 to a sudden resurgence through to 2013.

Moreover, taking a longer view, the study finds that when looking at the data from 1960, despite fluctuations, the world has on the whole become a far more unstable place, facing a dramatically higher scale of "political violence (PV) pandemics." As of 2013, the scale of "political violence pandemics" is three and a half times more intense than it was in 1960. Study author Henry Wilkinson writes: "Events such as the Arab Spring and, more recently, the wave of violent jihadist extremism affecting parts of the Middle East, have demonstrated the potential for individual outbreaks of unrest to trigger similar events across the world. These events generate widespread disruption yet prove extremely difficult to anticipate… the interdependencies which create the conditions for PV [political violence] pandemics are liable to become an increasingly important factor in determining international stability." The Lloyds study thus says that "instances of PV contagion (pandemics) have become more frequent, and the contagion effect ever more rapid and impactful" (Wilkinson 2016).

The Lloyds report argues that this increasing frequency and impact is linked to a range of trends such as the availability of the internet, urbanisation and geopolitical instabilities: "Growing complex interdependencies may make contagion one of the most important causal dynamics shaping how political violence emerges and spreads within and among states."

The report also identifies three fundamental "super-strains" of such PV pandemics, the first two of which are non-violent. The first consists of "anti-imperialist" and "independence movements", as well as social movements calling for the removal of an "occupying force."

The second consists simply of popular demonstrations, described as "mass pro-reform protests against national government." And the final "super-strain pandemic" is "armed insurrection" or "insurgency" and is associated with two particular ideologies, "Marxism" and "Islamism" (Wilkinson 2016).

The PV pandemic concept here is useful but the data is not without problems—Wilkinson mentions "Marxist" and "Islamist" inspired "super-strain pandemics", but makes no mention of the recent acceleration in Western military interventionism, nor the exponential rise in far-right extremism and terrorism.

6.3 Militarization

After 9/11, Western military interventions in the predominantly Muslim regions of MENA have escalated to an unprecedented degree. The data reveals that spikes in deaths from terrorist activity have increased directly following US military interventions in Afghanistan, Iraq and Syria. The vast majority of deaths from terrorism, the data reveals, have been Muslims occurring in just five countries: Iraq, Nigeria, Afghanistan, Pakistan and Syria. All the latter have been subject to direct and indirect forms of Western military and counter-terrorism assistance, with Nigeria having experienced the least degrees of such intervention (Fig. 6.4).

This of course underscores that geopolitical phenomena by themselves play a critical role in determining the dynamics of political violence. But it also reveals the centrality of the MENA region, where the vast majority of the world's fossil fuel resources are located, to the stability of the global geopolitical order and the neoliberal economy. Here, while the devastating impacts of Western interventions in the course of the 'war on terror' have clearly played a direct role in destabilizing key regions and fueling terrorism, this in turn has led to further Islamist militancy within MENA and beyond, which in turn has fueled far-right bigotry within the West against Muslim diaspora communities.

Yet a compelling body of literature—drawing on official documents, high-level government testimony, industry sources and statistical analysis—confirms that much of the military planning for the 'war on terror' preceded the events of 9/11.

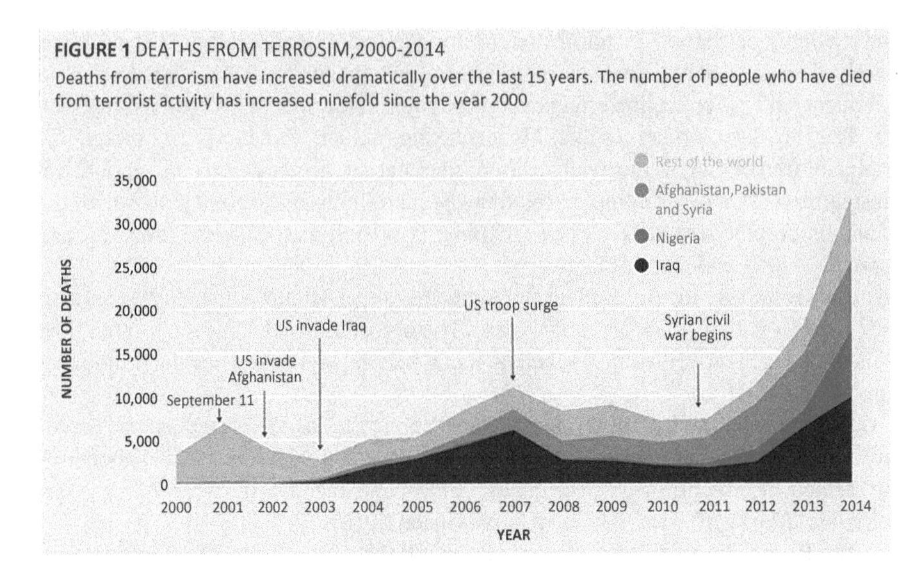

Fig. 6.4 Deaths from terrorism between 2000–2014. *Source*: ABC News (Australia) based on Global Terrorism Index

Major interventions in both Iraq and Afghanistan, and ongoing interventions across the region, whether direct, covert or based on indirect support for proxy forces, are invariably occurring in localities that are most strategically significant in relation to the world's remaining fossil fuel resources (Stokes and Raphael 2010).

This body of literature coheres with new scientific studies which confirm based on empirical data that US-led military interventions have been motivated by three principal concerns: access to hydrocarbon energy resources, mainly oil (Bove et al. 2015); forcing countries to open up their markets to US exports (Berger et al. 2013); and consolidating exploitative trade relationships favorable to US companies (Bove et al. 2014).

Explaining their findings on the role of oil in conflicts, Bove and Sekeris further observe: "We have shown that oil does motivate military interventions, and military assistance and interventions promote the commercial interests of the intervening country. This adds another dark dimension to the implications of modern societies' dependency on oil, while also raising questions about the ethical grounds of such military interventions" (Bove and Sekeris 2016).

This pattern of seeking dominance over the world's oil resources goes hand-in-hand with protecting the conditions for continued economic growth. Documents from one of the US military's most important commands, Central Command (CENTCOM), demonstrate that US military interventions have for decades been motivated to support the 'free market'. This has led to concepts of intertwined military and economic security being applied with the result that regions like the Middle East and Central Asia are seen as requiring "corrective military interventionism" to retain their subordination to the overarching CENTCOM mission of "keeping the global economy open" (Morrissey 2016).

This evidence underscores the role of Western military power in deploying means of violence as the principal mechanism by which to engineer socio-political relations in subject countries that are conducive to fossil fuel exploitation and integrating them into the global neoliberal capitalist order. Neither the centrality of oil to the global economy, nor the imperative of growth itself, can be abstracted from the essential function of state-led military-political violence in manufacturing the conditions in which the world's far-flung fossil fuel resources have been increasingly integrated into the global economy, often for the direct benefit of Western corporations, but more broadly with the result of sustaining global economic growth through increasing energy inputs.

There is thus a direct, symbiotic connection between the accelerating political violence of Western states in the form of military interventionism, and the political framework of market rules that underpins the global neoliberal architecture. As, however, EROI has continued to decline and geopolitical uncertainties have afflicted key states in MENA, this has led to an increasing resort to direct forms of Western military power projection to restore and maintain this global architecture, along with its regional and local pivots.

A 2013 RAND Corp analysis examined the frequency of US military interventions from 1949 to 2010 and came to the startling conclusion: not only that the overall frequency of US interventions has increased, but that intervention itself increases the probability of an ensuing cluster of interventions (Fig. 6.5).

The RAND report concludes that in one period, each new intervention "increases the likelihood of an additional intervention in the next by at least 20 percent and possibly as much as 50 percent." The report further found that such "clustered deployments" have been "more likely since the fall of the Soviet Union than during the Cold War," and further that the number of US interventions "increased dramatically over this period, especially between 1988 and 2004" (Kavanagh 2013).

Given the symbiotic link between US military interventions, energy interests, and the global economy, this establishes a strong empirical case for the conclusion that escalating Western state-militarization is a direct response to the destabilization of the global system as declining EROI has weakened the foundations of the global neoliberal capitalist order, and undermined state-territorial integrity in key strategic regions, particularly across parts of the Middle East, North Africa and Central Asia which contain most of the world's fossil fuel resources and energy transshipment routes.

6.4 Terrorism

In turn, the escalation of Western military interventionism has provoked an increase in Islamist militancy, which has further fueled far-right extremism, both comprising the principal sources of escalation in PV pandamics. Both, of course, have further elicited further militarization in response to these different forms of rising militancy and terrorism.

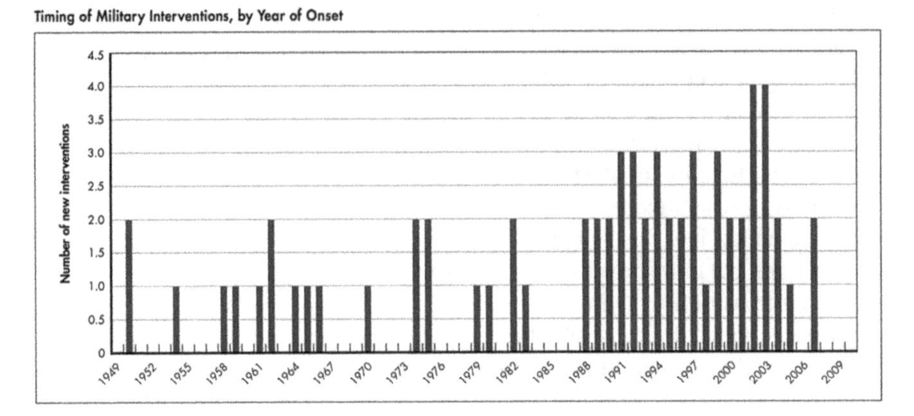

Fig. 6.5 Timing of military interventions by year of onset. *Source*: Kavanagh (2013)

Since the 1990s, the rise in Islamist militant groups has increased steadily, and the rate of increase has particularly accelerated in the period following the 2011 Arab Spring (Fig. 6.6).

Simultaneously, there has been a parallel rise in the number of far-right extremist groups in the US, coupled with an alarming rise in terrorist attacks by far-right groups worldwide. In the US, by far the biggest threat to homeland security is from far-right extremists, and within Europe, there has been a corresponding rise in popular support for far-right political parties (Fig. 6.7).

Research conducted by this author for the London-based hate crime charity Tell MAMA found that every single one of these far-right parties gaining popularity in Europe has strong neo-Nazi connections, yet mobilizes largely on an anti-Muslim platform (Ahmed 2016b).

There is therefore a direct correlation between the escalation of Islamist militancy and the rise of far-right extremism (Fig. 6.8).

The data shows that 2008 was a major spike year for far right attacks, while 2011 was a major spike year for Islamist militancy.

In both Islamist and far-right cases, the role of deeper systemic crises in triggering socio-political instability is apparent. The 2008 global financial crisis was partly triggered by the oil price hikes that year, which were driven by the plateau in conventional oil production since 2005. High prices fed into inflation and undermined the capacity of consumers to service their mortgage debts, thus playing a key role in the ensuing spate of defaults (Ahmed 2010).

In 2011, a major trigger of the Arab Spring uprisings were, in the wake of continuing oil price spikes, global food price shocks. Several studies confirm a direct link between the food price hikes and the outbreaks of political instability across the MENA region that followed. The global food price hikes were driven not just by the oil shocks, but more specifically by the global-level impacts of accelerating extreme weather events due to climate change, which in preceding years had led to a string of crop failures in major food basket regions, as well as local challenges due to

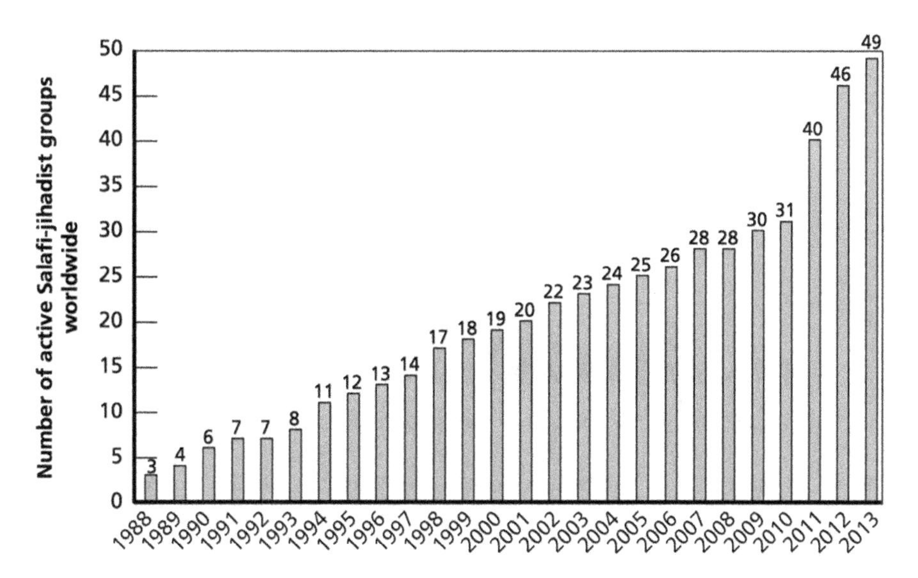

Fig. 6.6 Global rise of number of Islamist militant groups. *Source*: Seth G. Jones (2014)

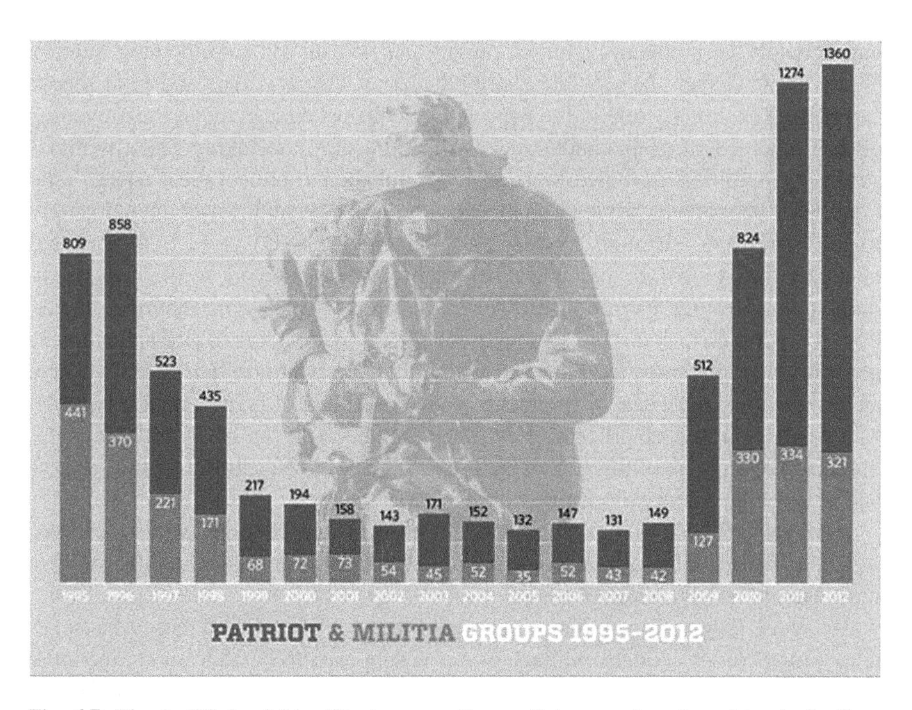

Fig. 6.7 Rise in US far-right militant groups. *Source*: Salon.com based on data via Southern Poverty Law Center

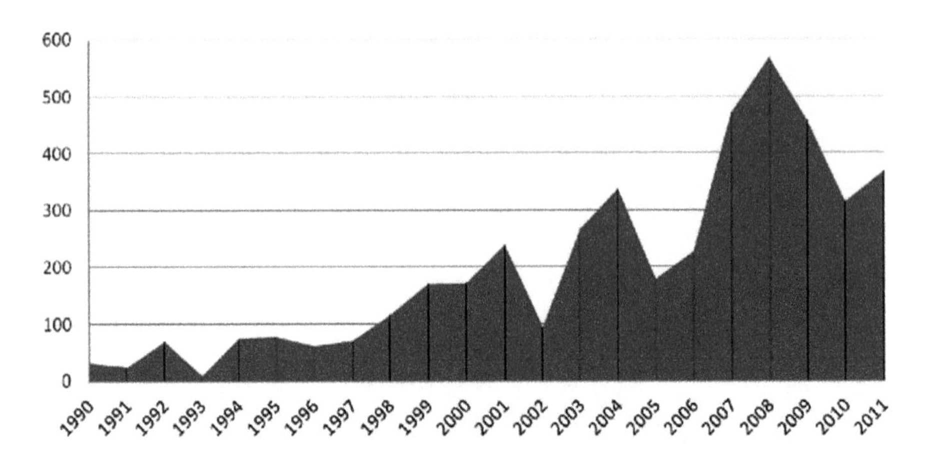

Fig. 6.8 Global rise in far-right attacks. *Source*: Arie Perliger, Combating Terrorism Center at West Point

political repression, mounting structural inequalities, massive dependencies on food imports, and a decline in local state revenues due to the inexorable depletion of national conventional reserves (Ahmed 2013a, c).

The analysis developed here thus allows us to theoretically frame these factors appropriately in preparation for empirical analysis in our country case studies. While 'triple crunch' factors like energy depletion, climate crisis and food shocks can be seen as directly relating to a growing global systemic crisis in the prevailing energy relations of global industrial civilization, they are taking place within a strained geopolitical framework originally established to protect these energy relations. ESD phenomena are leading increasingly to the breakdown of the old geopolitical framework—Human System Destabilization (HSD)—which in turn has elicited efforts to restore and adapt that geopolitical framework to the new conditions of accelerating disruption. However, the agents—states, international institutions, corporations, civil society groups, and so on—responding to HSD phenomena generally underestimate the extent to which they are being driven by ESD. These agents already have a deeply fragmentary understanding of ESD, which is seen not as a single overarching process of civilizational crisis driven by thermodynamics, but as a hodge-podge of forces that happen, somehow, to be connected, though how and why remains poorly conceived.

This leads us back to the role of Western military interventions, whose systemic function appears to be restoring or adapting the geopolitical framework declining in the wake of ESD-driven HSD.

However, judging by the intensification of the very phenomena that this accelerating projection of Western military power is supposed to be addressing, the latter has obviously not succeeded in this regard. Instead—and as will be specifically substantiated in ensuing chapters—it has destabilized the prevailing regional order, fueled Islamist terrorism, and indirectly thereby fueled a resurgence of far-right extremism motivated by anti-Muslim animus. This has led to an increase in state-militarization within Western homelands. Yet the heightened militarization at home and abroad has failed to stabilize this declining geopolitical order. Instead, it has accelerated its destabilization.

This reinforces the theoretical position articulated early on here, that the most powerful agents within global industrial civilization as we know it suffer from a fundamental knowledge-deficit. They fail to grasp the real nature of the global crises currently unfolding, their root structural causes and inherent systemic interconnections. Viewed from an evolutionary perspective, this means that as a complex adaptive system, key agents throughout global industrial civilization are pushing the system to a threshold of systemic crisis that threatens the very basis of the system's continued survival. The system must either adapt to these threshold effects by transforming its structure, adapting its overarching rules, norms and values, and thus transitioning to a new evolutionary state—or experiencing a protracted collapse process by failing to do so.

However, just as with biological organisms, the essential ingredient in the evolutionary process is the capacity to obtain and process information on environmental inputs accurately and then respond through genetic self-modification. Human civilization as a system must be able to do the same on a global-social scale. This brings us back to the central question of the critical role of knowledge-deficit in the acceleration of global systemic crises. The failure to diagnose, transmit and process information into knowledge that is actionable with a view to adaptation is the key lynchpin of the persistence of the global system in its current structure and its failure to adapt. The Global Media-Industrial Complex, representing the fragmented self-consciousness of human civilization, has served simply to allow the most powerful vested interests within the prevailing order to perpetuate themselves and their interests, rather than open up the way toward the global systemic transformation necessary to transition to a new evolutionary civilizational state.

In the following chapters, we will explore how the ESD-HSD model of global systemic crisis developed here can be applied to make sense of the patterns of a wide range of local and regional phenomena in recent years. Examining these empirical and historical case studies supplies us with important data by which to refine our model, and set out new lines of inquiry for further research by which to expand and test the model, as well as to identify the most productive pathways for civilizational transition.

Chapter 7
Biophysical Triggers of Crisis Convergence in the Middle East

The conventional narrative of the causes and consequences of the 2011 'Arab Spring' tends to focus on the idea of a democratic deficit in the region as the primary trigger, but fails to integrate this with a wider vision of the range of factors involved.

It is increasingly recognized that climate change played a major role in establishing conditions of societal vulnerability for the conflicts that followed the Arab Spring (Johnstone and Mazo 2011). Others argue correctly that the uprisings of the Arab Spring itself were triggered by unprecedented global food price spikes, (Lagi et al. 2011) while still others show that peak oil occurred in Egypt and Syria prior to the uprisings (Hallock et al. 2014). However, these studies neglect the systemic interconnections across these different factors. They thus fail to offer a truly systemic understanding of these phenomena.

In reality, the string of state failures across the region, and the inexorable swing toward multiple conflicts spurred on by the rise of various Islamist militant groups, can be traced directly to ESD (Earth System Disruption) phenomena unravelling the local sub-systems underpinning state integrity. In short, HSD (Human System Destabilization) in the form of the escalation of political violence has been fueled by ESD driven by interconnected biophysical processes of climate change, energy depletion and food crises.

7.1 From Syria to Iraq

The collapse of Syria into internecine warfare is, as with the Arab Spring, largely viewed as a direct consequence of the extreme political repression of President Bashar al-Assad, and the competing role of outside powers. To that extent, international policy has focused on viewing the conflict through the lens of geopolitical interests and regional security.

There has been some important recognition that climate change played at least an indirect role in catalyzing the Syrian conflict by creating a drought that led to social

N.M. Ahmed, *Failing States, Collapsing Systems*, SpringerBriefs in Energy,
DOI 10.1007/978-3-319-47816-6_7

pressures conducive to civil unrest. Yet there has been no recognition at all that a primary factor in the Syrian state's extreme vulnerability to such pressures was peak oil.

Prior to the onset of war, the Syrian state was experiencing declining oil revenues, driven by the peak of its conventional oil production in 1996 (Ahmed 2013b). Even before the war, the country's rate of oil production had plummeted by nearly half, from a peak of just under 610,000 barrels per day (bpd) to approximately 385,000 bpd in 2010 (Department of State 2014).

Since the war, production dropped further still, once again by about half, as rebels took control of key oil producing areas. Faced with dwindling profits from oil exports and a fiscal deficit, the government was forced to slash fuel subsidies in May 2008—which at the time consumed 15 % of GDP. The price of petrol tripled overnight, fueling pressure on food prices (IRIN 2008).

The crunch came in the context of an intensifying and increasingly regular drought cycle linked to climate change. A study in the *Proceedings of the National Academy of Sciences* has provided the most compelling research to date on how climate change amplified Syria's drought conditions, which in turn had a "catalytic effect" on civil unrest. The authors found that the 2007–2010 drought was the worst "in the instrumental record, causing widespread crop failure and a mass migration of farming families to urban centers", and was made three times more likely than by natural variability alone: "We conclude that human influences on the climate system are implicated in the current Syrian conflict" (Kelley et al. 2015). Compounding the impact of climate change, between 2002 and 2008, the country's total water resources dropped by half through both overuse and waste (Worth 2010).

Once self-sufficient in wheat, Syria has become increasingly dependent on increasingly costly grain imports, which rose by one million tonnes in 2011–2012, then rose again by nearly 30 % to about 4 million in 2012–2013. The drought ravaged Syria's farmlands, led to several crop failures, and drove hundreds of thousands of people from predominantly Sunni rural areas into coastal cities traditionally dominated by the Alawite minority. The exodus inflamed sectarian tensions rooted in Assad's longstanding favouritism of his Alawite sect—many members of which are relatives and tribal allies—over the Sunni majority (Agrimoney 2012).

Since 2001 in particular, Syrian politics was increasingly repressive even by regional standards, while Assad's focus on IMF-backed market reform escalated unemployment and inequality. The new economic policies undermined the rural Sunni poor while expanding the regime-linked private sector through a web of corrupt, government-backed joint ventures that empowered the Alawite military elite and a parasitic business aristocracy. Then from 2010 to 2011, the global price of wheat doubled—fueled by a combination of extreme weather events linked to climate change, oil price spikes and intensified speculation on food commodities—impacting on Syrian wheat imports. Assad's inability to maintain subsidies due to rapidly declining oil revenues worsened the situation (Friedman 2013).

As of 2010, Syria's then 20 million-strong population had one of the highest growth rates in the world, at around 2.4 %. In the first two months of 2011 alone, Syria's population grew by a monumental 80,000 people, most of whom were concentrated in the poorest eastern regions most badly affected by drought conditions (Sands 2011).

The food price hikes triggered the protests that evolved into armed rebellion, in response to Assad's indiscriminate violence against demonstrators. The rural town of Dara'a, hit by five prior years of drought and water scarcity with little relief from the government, was a focal point for the 2011 protests. The emerging Syrian conflict then paved the way for the rise of the Islamic State (ISIS) and other jihadist groups. Regional and international geopolitics fanned the flames of various rebel movements who moved into the widening vacuum of Syrian systemic state-failure to build new proto-state criminal enterprises.

Yet parallel processes have also been at play across the border, where ISIS is also active. US meteorologist Eric Holthaus specifically points out that the rapid rise of the Islamic State (ISIS) in 2014 coincided with a period of unprecedented heat in Iraq, recognized as being the warmest on record to date, from March to May 2014. Recurrent droughts and heavy rainstorms have also played havoc with Iraq's agriculture. With water supplies dwindling, and agriculture waning, Iraq's US-backed Shi'ite-dominated government has largely failed to address these burgeoning challenges, even as ISIS has moved quickly to exploit these failures, for instance by using dams as a weapon of war. Holthaus points out that climate-induced droughts have accompanied rapid population growth and agricultural stagnation, both of which are straining the capacity of the central government to feed its own population and deliver basic goods and services. As that state-level failure has been exacerbated, ISIS has rapidly filled the vacuum (Holthaus 2014).

Iraq's woes appear to have just begun. Figure 7.1 below demonstrates that on purely geophysical terms, Iraq's oil production is forecast to peak within a decade, by around 2025, before declining. However, the data shows that from 2001 onwards, Iraq's oil production has consistently remained below all three higher, middle and lower production scenarios due to above-ground geopolitical and economic complications. Such complications could usher in an effective peak earlier than anticipated.

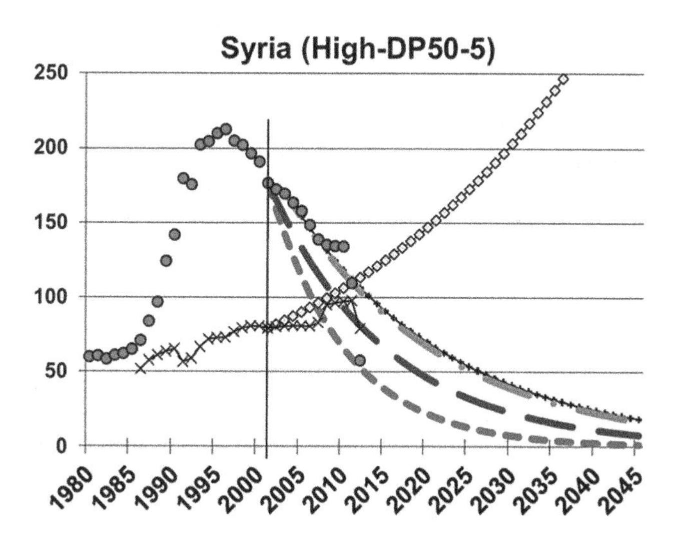

Fig. 7.1 Oil production forecasts for Syria (green = higher range; purple = medium range; blue = lower range; + = optimized forecast). *Source*: Hallock Jr. and Hall (2014)

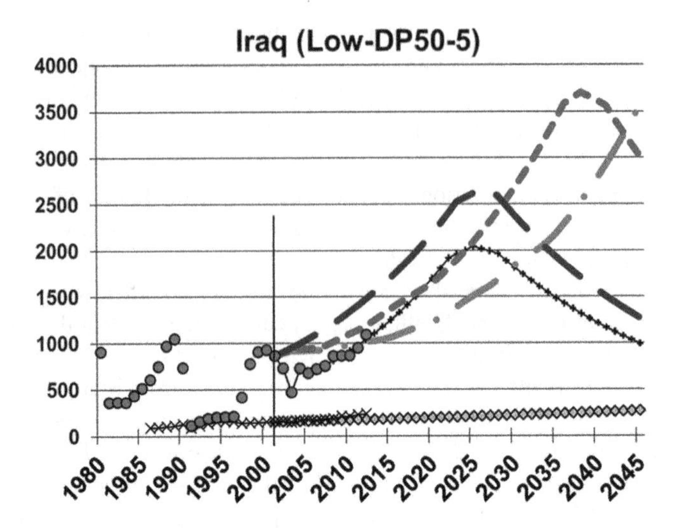

Fig. 7.2 Oil production forecasts for Iraq. *Source*: Hallock Jr. and Hall (2014)

Despite record oil production in April, the following month a senior official with Lukoil PJSC, operator of one of the country's biggest fields, told Bloomberg that crude output in Iraq "has probably peaked and is likely to fall short of the country's target over the next two years." Simply to maintain production at current levels, "Iraq needs more investment to maintain production," the official said (Dipaola 2016). To be sure, this unexpected brake on crude output is not primarily due to geophysical limits, but largely due to the escalating above-ground geopolitical turmoil in the region, combined with economic pressure from the declining profitability of the oil industry as market prices have bottomed out below production costs (Fig. 7.2).

The geopolitical crisis of ISIS and the 'war on terror' structure that is now in place across the region—which has evolved into a permanent geopolitical crisis-structure—should therefore be seen not as a purely political or religious phenomenon. While it certainly is both political and religious, this geopolitical crisis-structure constitutes a form of wide-scale Human System Destabilization that is a direct consequence of the cascading effects of deeper environmental, energy and economic crises across Syria and Iraq. These crises in turn are rooted in the escalating disruption of the Earth System due to the protracted failure of industrial capitalist subsystems in Syria and Iraq.

7.2 Yemen

The global conflict with ISIS playing out across Iraq and Syria is a mere taste of things to come, as well as a microcosm of processes playing out similarly across the wider region. Yemen is another major hot-spot for the 'war on terror', where the US and UK are purportedly fighting al-Qaeda in the Arabian Peninsula (AQAP), while simultaneously backing a Saudi Arabian military intervention to attack both AQAP but particularly the rival Iran-aligned Houthi clan.

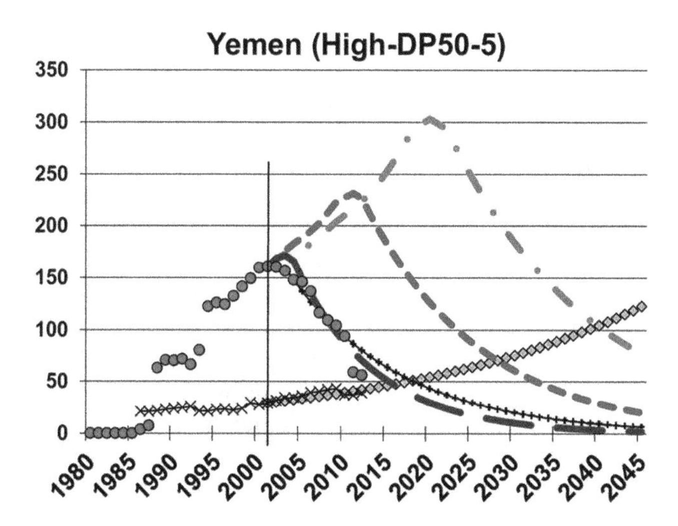

Fig. 7.3 Oil production forecasts for Yemen. *Source*: Hallock Jr. and Hall (2014)

Like Syria and Iran, Yemen has long faced serious water scarcity issues, which are being worsened by climate change. The country is consuming water far faster than it is being replenished, an issue that has been identified by numerous experts as playing a key background role in driving local inter-tribal and sectarian conflicts (Patrick 2015).

Yemen is one of the most water-scarce countries in the world. In 2012, the average Yemeni had access to just 140 m^3 of water a year for all uses, compared to the regional average of less than 1000 m^3—which is still well below adequate levels. Now in 2015, Yemenis have as little as 86 m^3 of renewable water sources left per person per year. The water situation in Yemen today is catastrophic by any reasonable standard. In many cities people have only sporadic access to running water—every other week or so. In coming years, Sanaa could become the first capital in the world to effectively run out of water (IRIN 2012).

Like Syria, Yemen is a post-peak oil country. Around 2001, Yemen's oil production reached its peak, since then declining from 450,000 barrels per day (bbd), to 259,000 bpd in 2010, and as of 2014 hitting 100,000 bpd. Production is expected to plummet to zero by 2017 (Boucek 2009) (Fig. 7.3).

This has led to a drastic decline in Yemen's oil exports, which has eaten into government revenues, 75 % of which had depended on oil exports. Oil revenues also account for 90 % of the government's foreign exchange reserves. The decline in post-peak Yemen state revenues has reduced the government's capacity to sustain even basic social investments. Things look grim now: but when the oil runs out, with no planning or investment in generating another meaningful source of government revenue, the capacity to sustain a viable state-structure will completely collapse.

At about 25 million people, Yemen has a relatively small population. But the population's rate of growth is exorbitantly high. More than half the population is under the age of 18 and by mid-century its size is expected to nearly double. In 2014, at a conference organized jointly by the National Population Council in Sanaa and the UN Population Fund, experts warned that within the next decade, these demographic trends would demolish the government's ability to meet the population's basic needs in education, health and other essential public services. But that warning is already transpiring. Over half the Yemeni population live below the poverty line, and unemployment is at 40 % generally, and 60 % for young people. Meanwhile, as these crises have fueled ongoing conflicts throughout the country, the resulting humanitarian crisis has affected some 15 million people (Qaed 2014).

A major impact of the high rate of population growth has been in the expansion of qat cultivation. With few economic opportunities, increasing numbers of Yemenis have turned to growing and selling the mild narcotic, which has accelerated water use to around 3.9 billion cubic metres (bcm), against a renewable water supply of just 2.5 bcm. The 1.4 bcm shortfall is being met by pumping water from underground water reserves. As these run dry, social tensions, local conflicts and even mass displacements are exacerbated, feeding into the dynamics of the wider sectarian and political conflicts between the government, the Houthis, southern separatists and al-Qaeda affiliated militants. This has also undermined food security. As around 40 % of Yemen's irrigated areas are devoted to qat, rain-fed agriculture has dropped by about 30 % since 1970. Like many other countries in the Middle East and North Africa, Yemen has thus become ever more dependent on food imports, and its economy increasingly vulnerable to global food price volatility. The country now imports over 85 % of its food, including 90 % of its wheat and all of its rice (World Bank 2014).

Between 2000 and 2008, the year of the global banking collapse, global food prices rose by 75 %, and wheat in particular by 200 %. Since then, food prices have fluctuated, but remained high. But rampant poverty means most Yemenis simply cannot afford these prices. In 2005, the World Bank estimated that Yemeni families spend an average of between 55 and 70 % of their incomes just on trying to obtain food, water and energy. And while 40 % of Yemeni households are in food-related debt as a result, most Yemenis are still hungry, with the rate of chronic malnutrition as high as 58 %, second only to Afghanistan (Arashi 2013).

For more than the last decade, then, Yemen has faced a convergence of energy, water and food crises intensified by climate change, accelerating the country's economic crisis in the form of ballooning debt, widening inequalities and the crumbling of basic public services.

Epidemic levels of government corruption, contributing to endemic levels of government mismanagement and incompetence, have meant that what little revenues the government has acquired have reportedly disappeared into Swiss bank accounts. Meanwhile, much-needed investments in new social programs, development of non-oil resources, and infrastructure improvements have languished. With revenues plummeting in the wake of the collapse of its oil industry, the government has been forced to slash subsidies while cranking up fuel and diesel prices. This has, in turn, cranked up prices of water, meat, fruits, vegetables and spices, leading to fuel and food riots (Mawry 2015).

As in Syria, the rise of violent and separatist movements across Yemen, including the emergence of al-Qaeda in the Arabian Peninsula (AQAP), has been largely enabled by the protracted collapse of the state. That process of HSD-level collapse has been driven primarily by ESD-level trends that are at play across the world, but manifesting locally: the peak of conventional oil production, intensifying extreme weather events due to climate change, the impacts on water and food scarcity, and deepening economic crisis.

7.3 Is Saudi Arabia Next?

One Gulf country that is especially vulnerable to this convergence of crisis within the next decade is Saudi Arabia. The kingdom's primary source of revenues, of course, is oil exports. For the last few years, the kingdom has pumped at record levels to sustain production, keeping global oil prices low, undermining competing oil producers around the world who cannot afford to stay in business at such tiny profit margins, and paving the way for long-term Saudi petro-dominance. But Saudi Arabia's spare capacity can last only so long. A recent study anticipates that Saudi Arabia will experience a peak in its oil production, followed by inexorable decline, in 2028 (Ebrahimi and Ghasabani 2015).

This could well underestimate the extent of the problem. According to the Export Land Model (ELM) created by Texas petroleum geologist Jeffrey J Brown and Dr. Sam Foucher, the key issue is not oil production alone, but the capacity to translate production into exports against rising rates of domestic consumption. Brown and Foucher showed that the inflection point to watch out for is when an oil producer can no longer increase the quantity of oil sales abroad because of the need to meet rising domestic energy demand. In 2008, they found that Saudi net oil exports had already begun declining as of 2006. They forecasted that this trend would continue (Brown and Foucher 2008).

They were right. From 2005 to 2015, Saudi net liquid petroleum exports have experienced an annual decline rate of 1.4 %, within the range predicted by Brown and Foucher. A report by Citigroup recently predicted that net exports would plummet to zero in the next 15 years. This means that Saudi state revenues, 80 % of which come from oil sales, are heading downwards, terminally (Daya and El Baltaji 2016). In this case, the forecast identified in Fig. 7.4 below—indicating a far lower production scenario than is widely recognized—might well put the peak date far earlier than 2028.

Saudi Arabia is the region's largest energy consumer, domestic demand having increased by 7.5 % over the last 5 years—driven largely by population growth. The total Saudi population is estimated to grow from 29 million people today to 37 million by 2030. As demographic expansion absorbs Saudi Arabia's energy production, the next decade is likely to see the country's oil exporting capacity evermore constrained. State revenues have been further hit through blowback from the kingdom's own short-sighted strategy to undermine competing oil producers. Saudi Arabia has maintained high production levels precisely to keep global oil prices low, making new ventures unprofitable for rivals such as the US shale gas industry and other

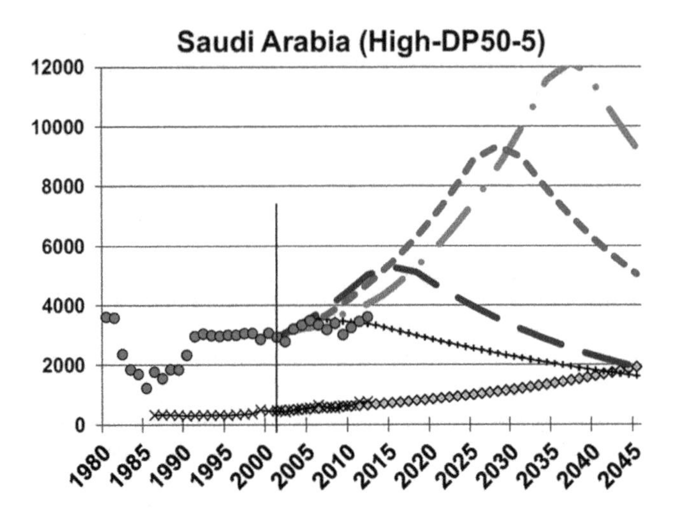

Fig. 7.4 Oil production forecasts for Saudi Arabia. *Source*: Hallock Jr. and Hall

OPEC producers. Yet the Saudi treasury has not escaped the fall-out from the resulting oil profit squeeze (Ahmed 2015).

The idea was that the kingdom's financial reserves would allow it to weather the storm until its rivals are forced out of the market, unable to cope with the chronic lack of profitability. Instead, Saudi Arabia's considerable financial reserves are being depleted at unprecedented levels, dropping from their August 2014 peak of $737–$672 billion in May—falling by about $12 billion a month. At this rate, by late 2018, the kingdom's reserves could deplete as low as $200 billion, an eventuality that would likely be anticipated by markets much earlier, triggering capital flight. To make up for this prospect, King Salman's approach has been to accelerate borrowing. What happens when over the next few years reserves deplete, debt increases, while oil revenues remain strained (Evans-Pritchard 2015)?

As with autocratic regional regimes like Syria and Yemen –which are facing various degrees of domestic unrest—one of the first expenditures to slash in hard times will be lavish domestic subsidies. In the former countries, successive subsidy reductions responding to the impacts of rocketing food and oil prices fed directly into the grievances that generated the "Arab Spring" uprisings. Saudi Arabia's oil wealth, and its unique ability to maintain generous subsidies for oil, housing, food and other consumer items, plays a major role in fending off that risk of civil unrest. Energy subsidies alone make up about a fifth of Saudi's gross domestic product. As revenues are increasingly strained, the kingdom's capacity to keep a lid on rising domestic dissent will falter, as has already happened in countries across the region (Peel 2013).

About a quarter of the Saudi population lives in poverty. Unemployment is at about 12 %, and affects mostly young people—30 % of whom are unemployed. Yet climate change is likely to heighten the country's economic problems, especially in relation to food and water. Like many countries in the region, Saudi Arabia is already experiencing the effects of climate change in the form of stronger warming temperatures in the interior, and vast areas of rainfall deficits in the north. By 2040, local average temperatures are expected to be higher than the global average, and could increase by as much as 4 °C, while rain reductions could worsen. This would be accompanied by more extreme weather events, like the 2010 Jeddah flooding caused by a year's worth of rain occurring within the course of just four hours. The combination could dramatically impact agricultural productivity, which is already facing challenges from overgrazing and unsustainable industrial agricultural practices leading to accelerated desertification (Chowdhury and Al-Zahrani 2013). In any case, 80 % of Saudi Arabia's food requirements are purchased through heavily subsidized imports, meaning that without the protection of those subsidies, the country would be heavily impacted by fluctuations in global food prices (Taha 2014). But with net oil revenues declining to zero—potentially within just 15 years—Saudi Arabia's capacity to finance continued food imports will be in question.

The kingdom is also one of the most water scarce in the world, at 98 m^3 per inhabitant per year. Most water withdrawal is from groundwater, 57 % of which is non-renewable, and 88 % of which goes to agriculture. In addition, desalination plants meet about 70 % of the kingdom's domestic water supplies. But desalination is very energy intensive, accounting for more than half of domestic oil consumption. As oil exports run down, along with state revenues, while domestic consumption increases, the kingdom's ability to use desalination to meet its water needs will decrease (Patrick 2015; Odhiambo 2016).

The confluence of ESD-level crises puts Saudi Arabia, and much of the rest of Arabian Gulf peninsula, in the firing line of HSD-level state-failure well within the next decade. Undoubtedly, if current forecasts of Saudi oil depletion are remotely accurate, then by 2030 the country will simply not exist as we know it. Coupled with the accelerating impacts of climate-induced water scarcity, and judging by the experience of some of its neighbors, there is no doubt that without a crash course to transition to a completely different energy system, the Kingdom is bound to begin experiencing systemic state-failure at most within 20 years, and probably much earlier.

7.4 The New Normal

While the Saudi case highlights that oil depletion will increasingly render the region's traditionally powerful oil exporting monoliths vulnerable, it also highlights the converging issue of climate-induced water scarcity. Research published by the American Water Works Association (AWWA) shows that water scarcity linked to climate change is now a global problem playing a direct role in aggravating major conflicts in the Middle East and North Africa. Numerous cities in Latin America, Africa, the Middle East, North Africa and South Asia are facing "short and declining water supplies per capita," which is impacting "worldwide" on food production, urban shortages, and even power generation. The UN defines a region as water stressed if the amount of renewable fresh water available per person per year is below 1700 m^3. Below 1000, the region is defined as experiencing water scarcity, and below 500 amounts to "absolute water scarcity". According to the AWWA study, countries already experiencing water stress or far worse include Egypt, Jordan, Turkey, Iraq, Israel, Syria, Yemen, India, China, and parts of the United States. Many, though not all, of these countries are experiencing protracted conflicts or civil unrest (Patrick 2015).

Citing the findings of the Gravity Recovery and Climate Experiment (GRACE) sponsored by NASA and the German Aerospace Centre, Patrick notes that between 2003 and 2009, the Tigris-Euphrates basin comprising Turkey, Syria, Iraq, and Western Iran "lost groundwater faster than any other place in the world except northern India". A total of 117 million acre-feet of stored freshwater was lost due to reduced rainfall and bad water management. If this trend continues, "trouble may be brewing" for the region (Patrick 2015).

The convergence of regional energy depletion and climate-driven water scarcity amounts to an ESD process that will, therefore, increasingly generate HSD by undermining the domestic power and territorial integrity of Middle East states, as well as the regional framework of geopolitical order.

The rapid empowerment of Islamist militant groups in recent years, exemplified in the emergence of an Islamist proto-state entity in the form of ISIS, is a direct consequence of an accelerating process of HSD manifesting principally in the form of a sequence of interconnected regional state-failures. While the 'war on terror' geopolitical crisis-structure constitutes a conventional 'security' response to the militarized symptoms of HSD (comprising the increase in regional Islamist militancy), it is failing to slow or even meaningfully address deeper ESD processes that have rendered traditional industrialized state power in these countries increasingly untenable. Instead, the three cases emphasized—Syria, Iraq and Yemen—illustrate that the regional geopolitical instability induced via HSD has itself hindered efforts to respond to deeper ESD processes, generating instability and stagnation across water, energy and food production industries.

ESD and HSD have thus become part of a regional self-reinforcing amplifying feedback process. Within each state, local ESD and HSD impacts fed back into each other over recent decades until they culminated in a threshold effect that tipped each country over into a period of systemic state-failure. The rise of ISIS demonstrates how these seemingly discrete episodes of national systemic state-failure contemporaneously and sequentially fed back into each other, creating a regional ESD-HSD amplifying feedback loop that has no choice but to worsen without transformative response intervening in the dynamics of sub-system failure, rather than reacting to its surface-symptoms and becoming part of the feedback loop itself.

These cases also demonstrate that Gulf powers like Saudi Arabia assumed to be largely resilient to regional crises—playing a role of interfering in these crises for their own interests—are not immune at all. The examples of Syria and Yemen show that in a context of ESD crisis convergence, it can take approximately 15 years from the date of local peaks of conventional oil production for the wide-ranging societal impacts of energy depletion to converge with the accelerating systemic pressures of climate-induced water and food scarcity, resulting in a significant outbreak of systemic state-failure.

Marin Katusa, chief energy strategist at Casey Research, reports that "conventional Middle East oil production is either in or approaching decline, even among the big hitters", and that "many Middle Eastern countries may stop exporting oil and gas altogether within the next few years, while some already have" (Katusa 2016). Katusa's concerns have been backed up by a model produced by oil analysts at Lux Research, who estimated that OPEC oil reserves may have been overstated by as much as 70 %. True OPEC reserves could be as low as 428.94 billion barrels, which could mean a global net export crunch as early as 2020 (Lazenby 2016).

Simultaneously, according to the Arab Organization for Agricultural Development (AOAD), the Middle East is experiencing a persistent shortage in farm products, a gap that has widened steadily over the last two decades. Across the region, food imports now run above $25 billion a year on a net basis. This has been driven by a rapid population growth since 1990 of nearly 2.34 % annually (MacDonald 2010). In other words, the period from 2020 to 2030 will see Middle East oil exporters experiencing a systemic convergence of energy and food crises. There can be little doubt, then, that on a business-as-usual trajectory, it is only a matter of time before the powerful oil exporting countries in the Middle East face an ESD-HSD convergence that will render regional state structures as we know it completely and permanently untenable.

Chapter 8
Biophysical Triggers of Crisis Convergence in Africa

The same fundamental ESD processes unfolding in the Middle East are also happening in Africa, with similar deleterious consequences in aggravating the societal potential for violent conflict. Here we examine two distinctive cases: Nigeria in West Africa, and Egypt in North Africa. In both cases, HSD is occurring in the form of the increasing destabilization of the state. This destabilization is being driven by ESD phenomena comprising interconnected environmental, energy and economic crises.

Conventional policy and media narratives focus on short-sighted counter-terrorism responses and other measures to shore-up state power in these countries—essentially a combination of military and financial empowerment designed to expand state-national territorial integrity, by increasing its monopoly in the means of violence, and advancing its capacity to deliver public goods and services using cheap debt-money via low interest financing.

Yet neither of these responses are capable of enhancing the state's resilience to ESD—only of enhancing the state's capacity to react to escalating symptoms of HSD. In doing so, an increasingly militarized response to HSD is incapable of ameliorating HSD—because such radicalized, militarized state responses are themselves forms of political violence that contribute to the escalation of HSD, and tend to elicit further forms of reactionary violence from local publics. Meanwhile, as the fundamental structural causes of escalating ESD remain unchecked, this paves the way for crisis convergence to occur yet again in the near future as ESD processes continue to accelerate with little amelioration.

8.1 Behind Boko Haram in Nigeria

As in the Middle East, lurking behind the emergence of Islamist militancy in Nigeria is the specter of climate change and its impacts on an already deeply unequal society. A study by the Congressionally-funded US Institute for Peace has confirmed a

N.M. Ahmed, *Failing States, Collapsing Systems*, SpringerBriefs in Energy,
DOI 10.1007/978-3-319-47816-6_8

"basic causal mechanism" that "links climate change with violence in Nigeria." The report concludes: "...poor responses to climatic shifts create shortages of resources such as land and water. Shortages are followed by negative secondary impacts, such as more sickness, hunger, and joblessness. Poor responses to these, in turn, open the door to conflict." Unfortunately, a business-as-usual scenario sees Nigeria's climate undergoing "growing shifts in temperature, rainfall, storms, and sea levels throughout the twenty-first century. Poor adaptive responses to these shifts could help fuel violent conflict in some areas of the country" (Sayne 2011).

According to the late Sabo Bako of Ahmadu Bello University, the 1980s "forerunner" to Boko Haram was the Maitatsine sect in northern Nigeria, whose members included many victims of ecological disasters leaving them in "a chaotic state of absolute poverty and social dislocation in search of food, water, shelter, jobs, and means of livelihood" (Sanders 2013). A year after the USIP study, *Africa Review* reported that many Boko Haram foot soldiers happen to be people displaced by severe drought and food shortages in neighboring Niger and Chad. Some 200,000 farmers and herdsman had lost their livelihoods and, facing starvation, crossed the border to Nigeria. "While a good number of these men were found in major cities like Lagos, pushing water carts and repatriating their earnings to the families they left behind", *Africa Review* reported, "others were believed to have been lured by the Boko Haram".

Indeed, one retired Nigerian security official told the journal that the Nigerian military recognized a correlation between regional climatic events, and an upsurge in extremist violence: "It has become a pattern; we saw it happen in 2006; it happened again in 2008 and in 2010. If you remember, President (Olusegun) Obasanjo had to deploy the military in 2006 to Yobe State, Borno State and Katsina State. These are some of the states bordering Niger Republic and today they are the hotbeds of the Boko Haram" (Mayah 2012).

Chukwuma Onyia, a former staffer in the Economic Community of West African States (ECOWAS), argues that the inadequacy of the government's climate adaptation programs led to "exposure of the vast population of farmers in northern Nigeria to harsh environmental effects, consequently generating conflict." He also highlights the intersection with other key factors: oil dependence and structural inequalities generated by neoliberal capitalist reforms: "Nigerian's over-dependency on crude oil rents, coupled with the behavior of political elites and the people-unfriendly market liberalizations foisted on Nigeria by the World Bank and IMF derailed the developmental focus of the state, and increasingly weakened its capacity to adapt to a changing climate, particularly in arid northern Nigeria." The failure to respond to Nigeria's changing climate generated increased drought and desertification, in turn leading to "decreased agricultural production, economic decline; population displacement and disruption of legitimized authoritative institutions and social relations." The net effect was an acceleration of the attractiveness of groups like "Boko Haram and other forms of Jihadi ideology," resulting in escalating "herder-farmer clashes emanating from the north since 1980s" (Onyia 2015).

Indeed, the rapid spread of Boko Haram coincided with the shrinking of the region's Lake Chad from 25,000 km^2 in 1963 to less than 2500 km^2, a phenomenon driven largely by climate change. At this rate, Lake Chad is set to dry up completely within 20 years. The disappearance of Lake Chad has led millions of people to lose their livelihoods. The result has been a groundswell of discontent, waiting for an outlet (Ahmed 2016a).

The other issue is Nigeria's intensifying energy crisis. In recent years, the country has intermittently faced fuel crises partly due to the government slashing previously high fuel subsidies, contributing to increasing public anger and civil unrest (Omisore 2014). According to one senior Shell official speaking off the record in 2014, crude oil production decline rates are "as high as 15–20%". Replacing this "natural production decline rate... requires more funds than is currently available". The same month, the head of Nigeria's petroleum resources department called for more investment in exploration to offset rapid decline rates: "Oil reserves are dropping, our output is dropping too... We need to do more in this regards to have more reserves. We have reached the plateau of production in the Niger Delta and we are already going down" (Ahmed 2014).

In the unlikely event that Nigeria resolves its internal turmoil, crude production might be increased to a plateau of around 3.5 mbd lasting a few decades. However, on conservative assumptions it is more likely that Nigeria's current plateau of 2.5 mbd could last just a few more years up to 2020 before declining from 36 % a year after 2025 (Dittmar 2016).

By early 2016, Wood McKenzie slashed its forecast for Nigeria's oil output for the next decade from 2.1 mbd to 1.5 mbd. Insiders revealed that about $15 billion of investment is required just to maintain current production levels and compensate for a natural decline in production of about 250,000 b/d each year. The industry narrative blames this predicament on corruption and ageing infrastructure, which surely play an important role—but they overlook the real elephant in the room: the end of cheap, easy oil. One study by two Nigerian scholars concluded in 2011 that "there is an imminent decline in Nigeria's oil reserve since peaking could have occurred or just about to occur; this is shown to be in agreement with previous studies" (Akuru and Okoro 2011). That conclusion could turn out to be premature if new investment, somehow, kicks in—but it is consistent with the production forecast of Hallock Jr. et al. (2014) as outlined in Fig. 8.1.

This verdict is corroborated by a 2013 report commissioned by the UK Department for International Development (DFID), which found that Nigeria's crude oil production has decreased since its peak in 2005, largely due to the impact of internal conflicts, leading to the withdrawal of oil companies and lack of investments. Since then production has fluctuated along a plateau. The DFID report noted that new offshore fields might bring additional oil on-stream, surpassing the 2005 peak—but also noted that rising domestic demand "at some point in the future may cut into the amount of oil available for export" (Hall et al. 2014). The drop in investment has been compounded by the new economics of unconventional oil: higher production costs, a supply glut due to the need for faster and more frequent drilling, and consequently lower market prices.

With Nigeria's population expected to rise from 160 to 250 million by 2025 and oil accounting for some 96 % of export revenue as well as 75 % of government

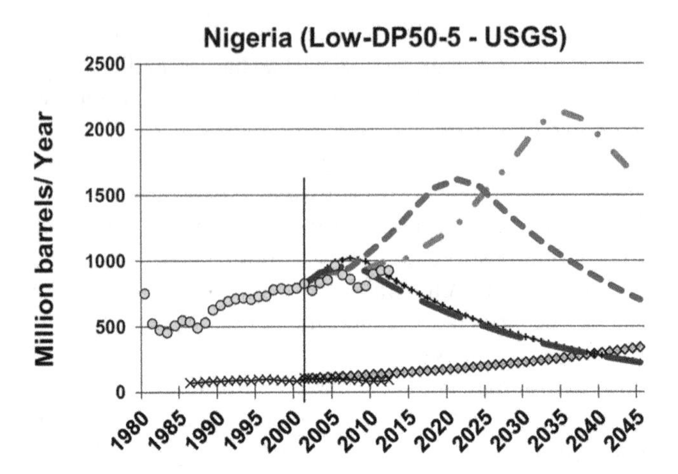

Fig. 8.1 Oil production forecasts for Nigeria. *Source*: Hallock Jr. et al. (2014)

revenue, the state has resorted to harsh austerity measures. Sharp reductions in public spending, power cuts, fuel shortages and conditional new loans are likely to exacerbate economic inequalities and further stoke the popular grievances that feed groups like Boko Haram in the North. With such domestic oil production challenges undermining Nigeria's oil export revenues, the fuel subsidy slash has pushed the brunt of the crisis onto the population, escalating the poverty and inequality that is a recruiting sergeant for Islamist terror.

In northern Nigeria, where Boko Haram hail from, about 70 % of the population subsist on less than a dollar a day. As noted by David Francis, one of the first western reporters to cover Boko Haram: "Most of the foot soldiers of Boko Haram aren't Muslim fanatics; they're poor kids who were turned against their corrupt country by a charismatic leader" (Francis 2014). Yet the biggest recruiting sergeant for this charismatic leader is the convergence of ESD crises of climate change, energy depletion, and agricultural decline.

8.2 Regime Rotation in Egypt

Just as the seemingly background factors of oil and climate played pivotal roles in laying the ground for the emergence of Boko Haram in Nigeria, the unsung villain of political turbulence in Egypt is the peak of its conventional oil production.

The major turning point for Egypt arrived in 1993, when the country's domestic oil production peaked at about 941,000 barrels per day (bpd), dropping since then to about 720,000 bpd in 2012, and further to 717,000 bdp in 2014 (also see Fig. 8.2). Similarly, domestic production of natural gas peaked at 6.06 billion cubic feet (bcf) of gas per day in 2009 (OBG 2016) Yet Egypt's domestic oil consumption has increased steadily over the past decade by about 3 % a year. Since 2010, oil

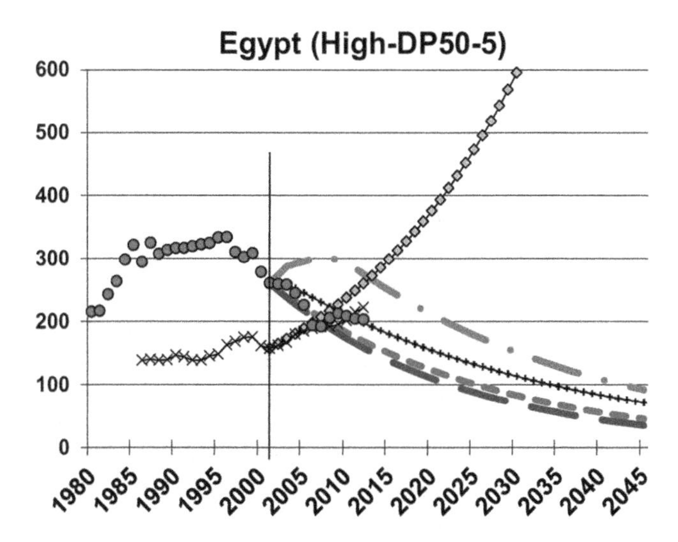

Fig. 8.2 Oil production forecasts for Egypt. *Source*: Hallock Jr. et al. (2014)

consumption—currently at 755,000 bpd—has outpaced production. It is no coincidence that the following year, Hosni Mubarak was toppled (Meighan 2016).

With Egypt's oil production well past its peak, its exports since 1996 have increasingly declined, despite inputs from new gas production. In 2009, as shown by data from BP's 2013 Statistical Review, oil exports had dropped by 26 %. According to Jean Laherrère, a petroleum geologist formerly with the French major Total S.A., two-thirds of Egypt's oil reserves have likely been depleted, and annual decline rates are already at around 3.4 % (Ahmed 2016a).

The impact on Egypt's state revenues has been dramatic. Energy subsidies amount to $15 billion a year, about a quarter of the entire budget, driven largely by expanding consumption needs for a growing domestic population. Over the last decade, Egypt's gas use has almost doubled, nearly matching production, further limiting the country's exporting capacity and, thus, hard currency revenues, reserves of which more than halved in two years (Kirkpatrick 2013a). In total, 10 % of its GDP is spent on subsidies (Coleman 2012).

With state revenues declining, how had Egypt experienced levels of growth of around 7 % in the two years preceding the 2008 global banking crisis—even winning praise from the World Bank, which described the government as a "top reformer"? The answer is simple: Egypt had financed increasing expenditures through one core mechanism: borrowing. Over the last decade, government debt has averaged about 86.77 % of GDP (TE 2016). In 2011, Egypt registered a balance of payments deficit of $18.3 billion. The situation became unsustainable as the state was increasingly unable to service myriad debts, had desperately attempted to identify viable sources of new oil and gas imports, but could not muster the capital to secure them (Shahine 2016). The heyday of growth, in other words, was achieved at a heavy socio-economic cost that was eventually paid on the streets in blood.

Increasing energy consumption, of course, is tied to an expanding population, which has grown exponentially by 21 % since 2000 to about 80 million people. This is projected to increase to about 100 million in the coming decade. Rapid population growth and continued economic mismanagement has meant that youth represent about 25 % of the population—but more than half of them suffer from poverty and unemployment (EI 2012). Economic mismanagement, much of which was quietly championed by the IMF and the World Bank, caused a widening of overall poverty while enriching mostly Egyptian elites.

As food subsidies declined in the context of declining state revenues, local food prices shot up. Once upon a time—in the 1960s—Egypt was completely self-sufficient in food production. Encouraged by international financial institutions to foster its export capacity, Egypt is now a net food importer, importing about 70 % of its food, and thus, vulnerable to global food price fluctuations (Saleh 2013).

Droughts and heat-waves in the US, Russia, and China since 2010 led to a sharp drop in wheat yields, on which Egypt is heavily dependent. That same year, Egypt's water shortages sparked tens of thousands of people to take to the streets in different parts of the country, primarily farmers protesting the growing inability to irrigate their farms—making tens of thousands of hectares of farmland impossible to cultivate (IRIN 2010). Egyptians in the 1960s enjoyed a water share per capita of 2800 m^3 (98,881 ft^3) for all purposes. The current share has dropped to 660 m^3 (23,307 ft^3)—well below the international standard defining water poverty at 1000 m^3 (35,314 ft^3) (Sarant 2013). Farmers have also been unable to afford to the price of diesel fuel used to power the pumps that irrigate their fields (Kirkpatrick 2013b).

In the run-up to the 2011 protests, global wheat prices doubled from $157/metric tonnes ($173/ton) in June 2010 to $326/metric tonne ($359/ton) in February 2011 (the same month Mubarak fell) while half the population was dependent on food rations. That year, the FAO Index averaged about 228, the highest since FAO started measuring international food prices in 1990. The second highest average occurred in 2008—the same year Egypt experienced violent clashes over government-subsidized bread in different cities, leading to 15 people being killed and 300 arrests. In May 2013, before Tahrir Square was flooded by millions of Egyptians, the food index was at 213 (UN News Center 2012).

While these crises have ameliorated, the risk that they could converge again is likely. The global food price index has stabilized, but Egypt's domestic food and electricity markets have not been fundamentally reformed, and continue to face the same structural problems. It is precisely the continued prevalence of those structural problems which meant that even after the overthrow of Mubarak, the replacement of his rule by the democratically-elected Muslim Brotherhood party led by Mohamed Morsi failed to ameliorate domestic grievances.

A month before Morsi himself was ousted in the wake of mass protests that led former army chief Abdul Fateh el-Sisi to oust the Brotherhood leader in a de facto coup, Egypt was experiencing a spike in urban inflation from 7.6 to 8.1 %, due to a weakening Egyptian pound—in the context of ongoing energy shortages and government austerity. Morsi was intent on securing a $4.8 billion IMF loan, and was further slashing energy subsidies while raising sales taxes to comply with the fund's neoliberal requirements. As Egypt's economic crisis made it harder to arrange payments, the country's wheat imports also dropped sharply to roughly a third of what it purchased a year ago. This, of course, simply exacerbated the HSD dynamics of the problem, and failed to address its root ESD causes.

The international community has looked to Egypt's new authoritarian ruler, el-Sisi, to stabilize the country with a combination of brute force, IMF loans, neoliberal reforms and continued hydrocarbon exploitation. These responses have apparently ameliorated the HSD trajectory, but *only temporarily*—they do not address the deeper ESD processes underlining the crises that triggered the 2011 chaos. El-Sisi's only real novelty is his willingness to use unprecedented levels of brutal force to crush opposition protests, to a degree even his authoritarian forbears had not done.

El-Sisi has, however, attempted to ramp up new investments in Egypt's unconventional energy resources, such as the offshore Zohr gas field, and it is possible that a production plateau could be maintained up to around 2021. However, this will then irreversibly decline by 3 % a year (Dittmar 2016). The predicament could well be even worse when this is squared with Egypt's export capacity relative to its domestic consumption requirements. Back in 2008, Hussain Abdallah, a former petroleum minister under Mubarak, forecast that Egypt's net oil and gas export capacity would terminate as early as 2020 due to the country's domestic energy demand (Al-Sinousi and Saleh 2008). The discovery of the Zohr field has offset this prospect for some years, and will allow Egypt to dramatically increase its gas production, with a view to serve both renewed exports and surging domestic demand. Yet it is unclear whether the project can retain sufficient profitability relative to production costs due to chronically low gas prices (Ghafar 2015).

It is equally unclear whether Egypt will be able to sustain the investments necessary to secure production of Zohr, without fatally denting its own ailing foreign currency reserves. Due to ongoing energy shortages, el-Sisi's government has been forced to ration electricity, leading to energy being cut for key industries. This, in turn, has widened economic losses, leading to declining exports and declining foreign reserves. This has decreased the state's ability to repay energy firms, leading to further decreased energy production, more gas shortages for industry, and increased government debt. That has reduced the value of the Egyptian pound, increasing import prices for manufactures, and disrupting energy for industrial production, further hampering exports, and once again reducing Egypt's foreign reserves. Egypt's energy and economy find themselves caught in an amplifying feedback loop (Barron 2016).

Zohr is forecast to reach 2.7 billion cubic feet per day in 2037, before declining to 1 billion cubic feet per day through 2045. This means that Egypt's supposed heyday of becoming 'energy independent' through gas—as the industry claims—may last a maximum of a decade at best (Adel 2016). Even if Zohr comes online, therefore, it will ameliorate the country's simmering energy crisis only temporarily. And by 2037, as Egypt's offset energy crisis returns, its climate-induced water and food crises will be far worse than they were in 2011. It is therefore only a matter of time before the existing framework of Egyptian political economic order that is attempting to clampdown on HSD is broken, as ESD processes of climate change, water scarcity, agricultural decline, and energy depletion continue to worsen.

8.3 Authoritarian Turn

Both Nigeria and Egypt demonstrate that the emergence of systemic state-failure due to the threshold effects of local ESD-HSD amplifying feedbacks can result in divergent outcomes depending on the local response. Rather than the states virtually collapsing into internecine civil war, they ultimately responded with extensive programs of domestic militarization justified under the rubric of counter-terrorism. However, such militarization obviously has no prospect of addressing the deeper ESD dynamics that generated the seeds of HSD in the first place. Yet it has been able to temporarily offset HSD.

Nigeria has had limited success with this offsetting, which means that as the corrupt mismanagement of its oil industry continues, it will be forced increasingly into a cycle of violence as its energy capacity continues to plateau amidst other converging crises.

In contrast, Egypt's Zohr discovery—hailed by the industry as a game-changer that could transform Egypt's economy through 'energy independence'—has merely permitted its new authoritarian ruler to offset HSD for about a decade. In reality, however, as this response remains very much part of a business-as-usual reaction failing to address deeper ESD processes, those process will continue to escalate, and it is unlikely that Egypt's intensifying water, food and economic crises will allow the country to remain stable within this decade. It does, however, increase el-Sisi's chances of offsetting full-blown systemic state-failure to the post-2030 period—after which, a local ESD-HSD amplifying feedback loop is likely to reappear with a vengeance.

Chapter 9
Biophysical Triggers of Crisis Convergence in Asia

Amidst the breakdown of the postwar geopolitical order in the Middle East and Africa, it is often presumed that the future for the 'Far East' is rosier. India and China are widely believed to be destined to become the economic powerhouses of the future, and the locus of a global economic shift to the East.

Whether or not such a shift takes place in any form, missing from this picture painted by pundits and politicians alike are the impacts of biophysical ESD processes which are already unravelling the Indian and Chinese development trajectories. The combination of energy depletion and climate crises, especially when seen in a global systemic context, have unavoidable ramifications that will within the next two decades bring both these economic juggernauts to the brink of systemic state-failure, without a dramatic transformation of the political economies of their energy production and consumption.

While neither India nor China are currently engaged in major internal conflicts that threaten to undermine national state integrity, the backdrop of crisis convergence in the Middle East and Africa suggests that this is likely to be in store in coming decades on a business-as-usual trajectory—although the pathway toward system threshold effects triggering prolonged state-failure in economies that import energy will necessarily be quite different.

9.1 The End of the Indian dream

According to the International Energy Agency's (IEA) projections out to 2040, Indian energy demand will grow more than any other country, more than doubling, and accounting for 25 % of the rise in global energy use to 2040, and the largest absolute growth in both coal and oil consumption. Even so, the IEA concedes that this massive increase in domestic energy production, based largely on oil, coal and natural gas (as well as renewables) will remain far below India's consumption needs, requiring more than 40 % of primary energy supply to be imported by 2040,

N.M. Ahmed, *Failing States, Collapsing Systems*, SpringerBriefs in Energy,
DOI 10.1007/978-3-319-47816-6_9

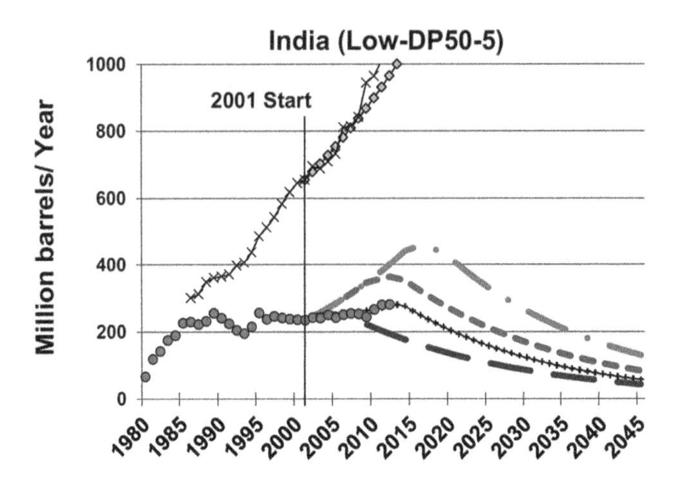

Fig. 9.1 Oil production forecasts for India. *Source*: Hallock Jr. et al. (2014)

up from 32 % in 2013. The IEA also concedes that India's oil production, which is already tepid, will decline further resulting in a rapid rise in net oil imports, increasing the country's oil import dependence from its current level of 80 % to over 90 %, especially from Middle East exporters. This, however, is precisely the period in which Middle East exports are expected to dramatically decline as the major oil exporters begin to scale the limits of their oil export capacities (IEA 2015).

India's chronic dependence on energy imports is unavoidable. As Fig. 9.1 illustrates, India is on the brink of entering into terminal decline of its domestic oil production. Despite having about 17 % of the world's population, India holds only 0.3 % of the world's proved reserves of oil; 0.7 % of the world's proved reserves of gas; 7 % of the world's proved reserves of coal; and 2 % of the world's uranium. The IEA's projections demonstrate that even if India rapidly mobilizes its renewable energy potential, it will not be able to do so sufficiently to meet its own rocketing energy consumption requirements driven by rising industrial economic demand, without relying increasingly on fossil fuels. Indeed, if India was to rely only on its own renewable energy resources, the country would use a massive 73 % of its entire renewable potential in just a decade (Tranum 2013).

Unfortunately, this increases the temptation for India to ramp up its domestic coal production, which would have extremely dire environmental consequences—not least for India itself. Although according to Goldman Sachs world coal production is at or near peak, India retains significant potential to continue growing production (Cunningham 2016). One study puts the date for peak coal in India at 2059—and while this would provide ample scope to burn enough coal to help trigger dangerous global warming, it would not keep up with the pace of projected demand growth in India (Nandi 2014).

Data from BP's 2016 *Statistical Review of World Energy* confirms that since 1996 there has been an increasing gap between India's production and consumption of coal, such that by 2014, India's own coal production fell far short of domestic consumption—a gap that continued to widen up to 2016 (Tverberg 2016). Only by

2016 did India's coal imports undergo a marked 15 % decline since the preceding year, forecast to continue to 2017—as domestic coal production increased. However, it is unclear whether this trend can continue, even as it is clear that coal alone cannot meet India's overall projected energy demand (Buckley 2016).

India faces not just a looming energy crisis, but also a converging water and food crisis, radicalized by climate change. In 2016, India experienced its worst water crisis on record, with a quarter of its population suffering from a second successive drought, leading to crop failures. India's population is forecast to rise by a further 450 million by 2050. Yet according to the Water Resources Group, India's national water supply is forecast to fall 50 % below demand as early as 2030 (Timms 2016).

So far, the Indian government's measures to prepare for these crises have been woefully inadequate. Climate change will lead to increasing crop failures, and escalating food prices, which in turn will have debilitating impacts on rural economies. India's rapid population growth, which is hoped to be a potential source of future economic growth in which rising domestic demand provides a robust internal market—a 'demographic dividend'—would end up acting as a brake on economic growth due to the increasing inability to meet domestic demand under environmental crises. What is hoped to be a 'demographic dividend' would end up becoming a demographic disaster (Pritchard 2016). As of 2015, India has an estimated population of 1.31 billion people, which is rising at a rate of 1.22 %. By 2022, India is projected to overtake China's population, becoming the most populous country in the world with 1.4 billion. This would continue to rise to 1.5 billion by 2030, and 1.7 billion by mid-century (UN 2015).

According to a UK Ministry of Defence report projecting the region's future out to 2040, climate change will lead to "rising sea levels … increased incidents of seasonal floods, heat-waves, storms, and unpredictable farm yields." If sea levels rise quicker than anticipated, "millions of people across South Asia (principally in Sri Lanka, Bangladesh and the Maldives) will be displaced, with no opportunity to return to their homes." Irregularities in the pattern of monsoon rains are likely to undermine South Asia's "agricultural and domestic water needs", while higher temperatures will "increase the range of vector-borne diseases such as malaria", such that it becomes "prevalent all-year-round" (DCDC 2013).

Yet such impacts are not merely slated for the seemingly distant future, they are already happening. Two consecutive droughts for the first time in three decades in 2016 has led to dramatically slashed agricultural output for India's staple food exports, wheat, corn and oilseeds. India already has severe trouble feeding its domestic population, where 42 % of children are malnourished—largely because the poor cannot afford to buy food grains. The agricultural decline meant that in 2016, India began to import corn simply to bring down local prices (Bhardwaj 2016). Official statistics, which have been heavily critiqued as being politized, put national poverty at some 22 %, declining from 45 % in 1993. However, the Global Consumption and Income Project, adopting a more robust poverty line, finds the rate of poverty in India to now be 47 %, nearly half the total population (Kundu 2016).

As climate change kicks in, water and food scarcity will increasingly hit India's state revenues, exacerbating agricultural decline, while fueling local price hikes and increasing India's food import dependency. Either way, India will find itself caught between local price spikes and declining state revenues due to increased expenditures on imports.

India's high rate of economic growth has thus not only failed to 'trickle down' to the majority of the Indian population, it has come at a huge environmental cost. In 2008, the Chamber of Indian Industries conceded that India was already consuming twice as much as its natural resources can sustain. As Kothari reports: "Net growth in employment in the formal sector has been insignificant, with over 100 million people having to find desperate avenues of livelihood in the informal and often exploitative economy. Of course there is much more infrastructure, many more industries, and hundreds of more shopping malls, but over two-thirds of Indians remain deprived of one or more of the basic needs: adequate nutritious food, clean water and air, appropriate shelter and sanitation, energy, opportunities for learning and good health, and productive livelihoods" (Kothari 2014).

Indian economist Prasenjit Bose challenges official statistics on India's high growth rates, noting that they appear to fudge countervailing figures from India's 2016 economic survey, indicating that gross fixed capital formation (investment) has fallen from 33.4 % of GDP in 2012–2013, to 30.8 % in 2014–2015, and further to 29.4 % in 2015–2016; agriculture has grown by a paltry 1.1 % in 2016, with food grain production stagnating at around 250 million tons for the past two years; exports and imports fell by 17.6 % and 15.5 % respectively in 2016; and inflation remains around 5 % even though Indian basket crude oil prices fell from $84 to $30 per barrel: "These are certainly not the signs of a booming economy" (Bose 2016).

These underlying fault lines will begin to crack open when India's booming energy requirements—requiring increasing state expenditures—hit the lowering ceiling of water and food scarcity due to climate change and higher energy costs.

The pathway toward a crisis convergence that could lead to protracted systemic state-failure will therefore be fundamentally distinctive to that experienced in present or former oil exporting countries in the Middle East. As India is historically a net energy importer with minimal domestic fossil fuel resources, the growth of its economy has always been structurally dependent on its ability to access cheap energy from abroad. Hence, the threshold for systemic state-failure will depend on the convergence point between global peak oil, the peak of India's domestic oil resources, an accelerating decline in high quality energy imports from the Middle East, and the extent to which this decline undercuts rapidly rising Indian economic demand due to population growth. That convergence point in turn will be accelerated by the myriad societal impacts of climate change, particularly on water and food scarcity. India's predicament is hardly unique for the region. Its neighbor, Pakistan, faces similar challenges (Pracha and Volk 2011).

9.2 China: Paper Tiger

Unlike India, China is a far more significant oil producer. But a comprehensive study finds that China's conventional oil production has already peaked in 2010. Increasing production is therefore shifting toward China's remaining unconventional resources.

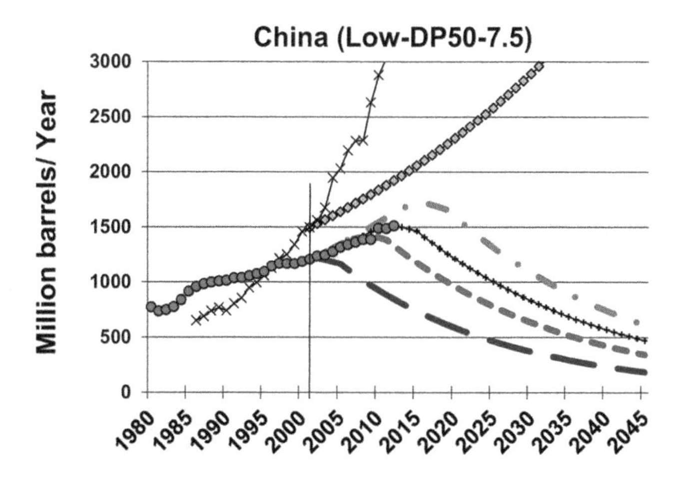

Fig. 9.2 Oil production forecasts for China. *Source*: Hallock Jr. et al. (2014)

However, the study warns that unconventional production "will be especially adversely affected by low prices" and therefore could become increasingly unprofitable. Total conventional and unconventional oil production is likely to peak around 2020 (K. Wang et al. 2016) (Fig. 9.2).

China is also facing peak coal. While many projections have put the date of a peak in China's domestic coal production between 2020 and 2040, a *Nature Geoscience* analysis argues that a more accurate accounting of available data indicates that China's peak coal point has already passed in the year 2014 (Qi et al. 2016).

As with India, China is already embarking on an ambitious investment plan for renewable energy, but it is widely recognized that the current commitment is subpar, and cannot possibly make-up for the decline in fossil fuels. China has pledged to increase the share of clean energy sources in its primary energy mix to 15 % by 2020, and to 20 % by 2030. Despite astronomical investments, growing domestic demand amidst a fast growing Chinese population means that the remainder will be made-up from China's depleting fossil fuel resources, and rising oil and gas imports from the Middle East. Domestic demand will be driven by China's large population. Although the rate of population growth is already slowing—its current population of 1.381 billion people is projected to rise but then mildly plateau and decline up to 2050, stabilizing around the 1.38 billion mark (Street et al. 2014)—China's working middle class population is forecast to increase by another 100 million people by 2030. Consumer spending by this class is expected to more than double over the same period (ICEF 2016).

Yet like India, China faces the problem that as we near 2030, net exports from the Middle East will track toward zero at an accelerating rate. Precisely at the point when India and China's economic growth is projected to require significantly higher imports of oil from the Middle East, due to their own rising domestic energy consumption requirement, these critical energy sources will become increasingly unavailable on global markets. This date would thus appear to provide a clear crunch-point at which Indian and Chinese economic growth would not only face an impassable ceiling, but where the threshold effects of system failure could begin to generate state-failure due to the impact of energy scarcity in terms of declining state revenues. Chinese economist Minq Li of the University of Utah thus forecasts that: "China's economic growth rates are projected to fall steadily from 2015 to 2050. After 2030, China's economic growth rate will fall below 3.5 percent… After 2050, the Chinese economy will enter into a quasi-steady state as the economic growth rates fluctuate around zero" (Li 2014).

This declining rate of economic growth will be exacerbated by climate change according to a study published by the World Health Organization. Projections show that global warming along with land conversion and water scarcity could reduce Chinese food production substantially in coming decades. Climate change could reduce China's per-capita cereal production, compared with that recorded in 2000, by 18 % by the 2040s. By 2030–2050, loss of cropland resulting from further urbanization and soil degradation could lead to a 13–18 % decrease in China's food production capacity—compared with that recorded in 2005. These declines could result in "continued or recurring food shortages" posing a "substantial threat to overall community health and well-being, social stability and human nutrition" (Tong et al. 2016).

Earlier studies have been more ambiguous on the potential climate impacts on Chinese agriculture. One using survey data of 8405 households across 28 provinces found that irrigated farms could benefit from climate change, but that rainfed farms would be increasingly damaged, especially in the Northeast and Northwest. (J. Wang et al. 2009) However, more recent studies like the above from the WHO have highlighted the risks more precisely. They confirm that the major risk to Chinese agriculture will be in the North, from "water-related challenges in coming decades due to the expected increases in water demands and soil-moisture deficit, and decreases in precipitation" (Tao et al. 2003).

China is a major importer of food grain, and climate change is likely to increase the gap between its domestic consumption needs and national food production. In 2013, the FAO forecasted that China's imports of coarse grains to feed livestock would double by 2022. Soyabean imports would grow 40 %, and imports of beef would nearly double: "The challenge is clear: feeding China in the context of its rapid economic growth and limited resource constraints is a daunting task. China's consumption growth will slightly outpace its production growth" (Hook 2013). In fact, climate change has already caused economic losses of about $820 million to China's corn and soybean crops in the past decades. In the long-term, climate change is projected to lead corn and soybean yields to decline by 3–12 % and 7–19 % respectively, by end of century (Chen et al. 2016).

China is thus forecast to dramatically increase both its food and energy import needs, even while geophysical and climate realities will exert a brake on China's domestic energy, water and food production. As China's increasingly expensive energy requirements slow down its capacity for economic growth, the state will be forced to dramatically uptake its energy imports at a time when it cannot afford to, and when global energy markets will be even tighter than they are now. Moreover, while China's domestic coal production has already likely peaked, China's conventional oil supplies have already peaked, and when production from China's total conventional and unconventional oil supplies are examined together, their peak is less than five years away.

9.3 Arrested Development

For both India and China, then, the prospect of mimicking the historical industrial development trajectories of the Western world is a fantasy that ignores the ground reality of ESD. The myopic effort to follow the same model of growth is accelerating ESD processes whose impacts are set to unravel South Asia's regional political and economic order well within the next two decades. HSD will begin to accelerate within this period in the region in the form of widening outbreaks of civil unrest. Depending on how the Indian and Chinese states respond, it is likely that these outbreaks of domestic disorder will become more organized, and will eventually undermine state territorial integrity before 2030.

This is not a foregone conclusion, but it is the most likely outcome of a business-as-usual scenario in which India and China continue to pursue policies driven by their capitalist endless growth imperative. While India's economy is forecasted to grow in the near-term, its continued growth will come at the expense of its own biophysical basis and environmental health, eventually undermining itself. Similarly, China is on the brink of entering a new era of domestic energy scarcity. Both will therefore become increasingly vulnerable to climate-induced water and food crises.

The fabled shift of core power to the East, therefore, is unlikely to happen in the way envisaged by some, simply because the Indo-Chinese developmental trajectory in its current form is destined to become increasingly cannibalistic. For these economies to continue growing will require an ever increasing consumption of resources in terms of food, water and energy. Yet the very intensification of the exploitation of these resources to fuel growth will increase the costs of exploiting and consuming them as net energy imports become more expensive, decline, and as fossil fuel consumption accelerates the destabilizing impacts of climate change. In this way, the very foundations of economic growth in India and China will, as we approach 2030, increasingly accelerate the environmental and energetic costs of that very growth, in turn undermining growth itself.

Chapter 10
Biophysical Triggers of Crisis Convergence in the Euro-Atlantic Core

The uncertain fates of India and China raise awkward questions about developments from Russia, to Europe, to America, where energy depletion is either rapidly underway, or imminent within the next decade.

Within Europe, resource depletion has meant that the European Union as a whole has become increasingly dependent on energy imports from Russia, the Middle East, Central Asia and Africa. Yet exports from these regions will become tighter as major oil producers approach production limits.

Further, geopolitical turmoil that has unfolded in Ukraine provides a compelling indication that such HSD processes are rapidly moving from the periphery of the global system into the core. For the most part, the Euro-Atlantic core—traditionally representing the most powerful sections of the world system—has insulated itself from global crisis convergence impacts by diversifying energy supply sources. However, there is only so much that diversification can achieve when the total energetic and economic quality of global hydrocarbon resource production is declining. In the next decade, the Euro-Atlantic core will be forced to contend with this reality as oil exporters in the periphery of the world system breach their production limits and see their net exports begin to plateau and decline.

10.1 Europe

Europe is now well into the throes of a post-peak oil scenario (see Figs. 10.1, 10.2, 10.3 and 10.4). A study commissioned by the former French environment minister Yves Cochet, published by the Greens-European Free Alliance Group in the European Parliament, notes that the EU is the second largest consumer of oil in the world after the US. Oil accounts for 38 % of primary energy consumed in the EU, and by 2020, dependence on energy imports will rise to 92 % of total requirements. Meanwhile, the quantity of oil produced within the EU is declining. Around 2000, Europe produced up to 25 % of its own oil, but today this has declined to 13 %.

N.M. Ahmed, *Failing States, Collapsing Systems*, SpringerBriefs in Energy,
DOI 10.1007/978-3-319-47816-6_10

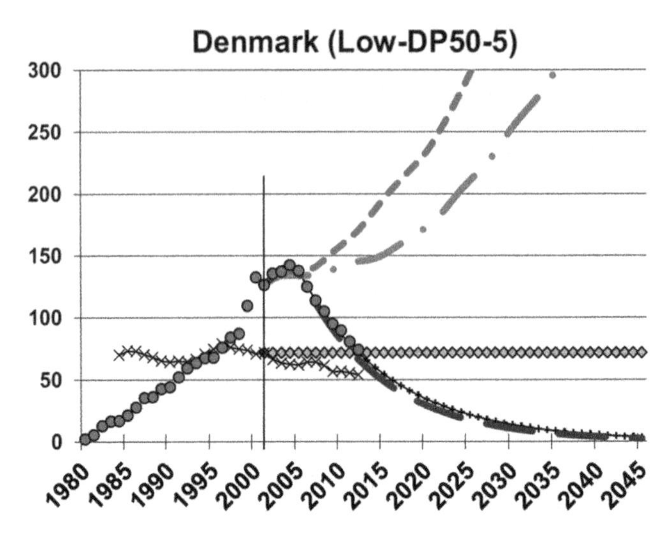

Fig. 10.1 Oil production forecasts for Denmark. *Source*: Hallock Jr. et al. (2014)

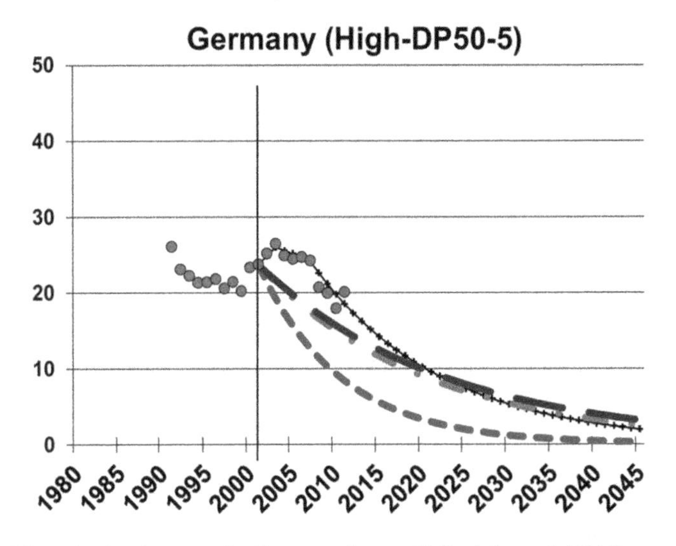

Fig. 10.2 Oil production forecasts for Germany. *Source*: Hallock Jr. et al. (2014)

Europe currently produces 1.7 mbd of conventional oil, and production is falling at the rate of 6 % per year since 1999. This is because all of Europe's major domestic producers are in decline: "Two-thirds of the oil produced is supplied by the United Kingdom, which is already past its peak oil production point and has been a net importer of oil since 2005. Denmark, the EU's second largest oil-producing country, passed peak oil in 2004 and its production is declining at the rate of 8–10 % per year, while Italy and Germany produce only 10 % and 5 % of their requirements respectively. For its part, Romania passed peak oil in 1973. European oil-producing

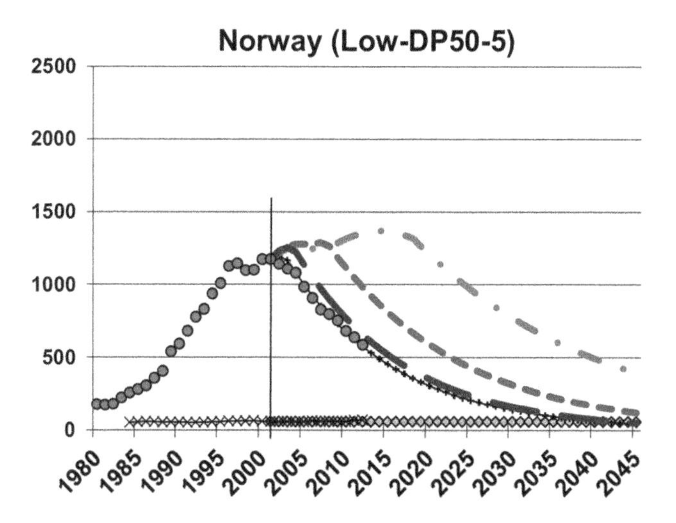

Fig. 10.3 Oil production forecasts for Norway. *Source*: Hallock Jr. et al. (2014)

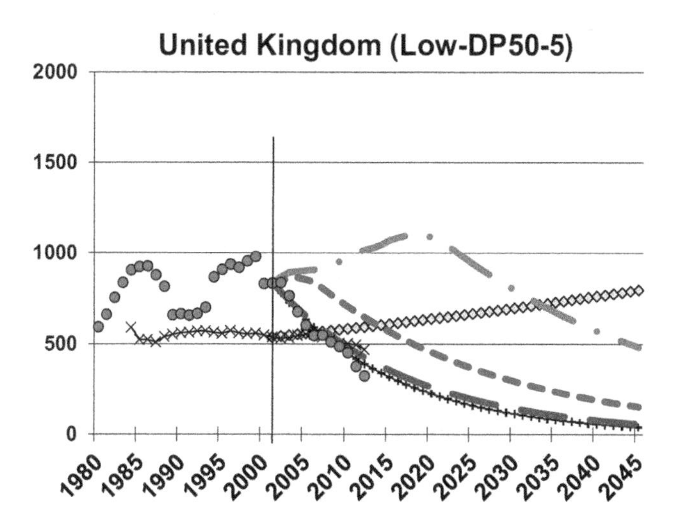

Fig. 10.4 Oil production forecasts for UK. *Source*: Hallock Jr. et al. (2014)

countries have all passed peak oil, and only Denmark produces more oil than it consumes" (Thevard 2012).

As with North America, there are hopes that Europe's shale gas resources could help increase domestic self-sufficiency in energy production. However, a report by the European Academies Science Advisory Council (EACAS) concludes that: "… the scale of the resource itself and the economic viability of its extractions in different Member States remain uncertain", and will remain so without sufficient exploratory drilling (EASAC 2014).

So far, exploratory drilling has confirmed that shale gas's capacity to offset the decline in conventional production is severely limited: "The exploratory results up until now have been disappointing," finds one study examining case studies in Poland, the UK and Germany, "which appear to be close to shale gas production, but to the extent of making them energy self-sufficient, or secure in their energy needs, that will never be realized" (Dodge 2016). Shale gas plays in Sweden and Turkey have also been found to lack economic feasibility (Weijermars 2013).

While there remains considerable uncertainty over the prospects for shale gas in Europe, Russia, which is the leading supplier to the EU, is on the brink of its own production peak. As Michael Dittmar of the ETH Zurich Institute for Particle Physics argues, "neither Russia nor the other oil exporting countries from the FSU [former Soviet Union] will be able to replace the missing million barrels of oil during the next decades." Russia's oil production has plateaued for the last few years at about 10 mbd, mostly due to the decline of Russia's largest fields. Even accounting for the rise in production from newer fields in Eastern Siberia, the prospects are limited. Western Russia and Siberia have begun a 3 % a year decline phase, which will accelerate to 6 % per year from 2020 onwards. In Eastern Siberia, production is forecast to rise nominally over the next few years, then plateau at around 1.6 mbd from 2020 to 2025, before declining "by 3 %/year to 2030 and 6 %/year afterwards" (Dittmar 2016).

If Europe is hoping that exports from the key Central Asian producers, Azerbaijan and Kazakhstan, will help—they will not. Production from Azerbaijan has declined by 15 % from 2011 to 2014. This is likely to accelerate to 6 % a year after 2020. Kazakhstan is forecasted to hit a production plateau imminently in 2018, lasting until 2023, after which it will decline 3 % per year and a further 6 % a year after 2028. Taking into account rising domestic consumption of oil resources, this means that total oil exports from the FSU to Western Europe have already begun declining, from 6 mbd in 2014 to about 4 mbd (2020), "and will essentially be terminated around the year 2030" (Dittmar 2016).

Within the same period, Saudi Arabian net oil exports have been forecast to decline to zero, which means that the EU faces not simply an ongoing decline in its domestic production of liquid fuels, but is simultaneously experiencing a decline in imports, which will be mirrored in other parts of the world.

As we approach 2030, climate change will increasingly disrupt European agriculture. While crop yields are forecast to grow in northern Europe, in southern Europe and the Mediterranean water shortages and extreme weather will lead to decreased yields, exacerbating the North-South divide within Europe. (Bindi and Olesen 2010)

However, the biggest pressure on Europe will come from outside. By 2030–2040, vast swathes of the Middle East and North Africa will become uninhabitable due to searing surface temperatures under local impacts of global warming. As crisis convergence unravels the global food system across the Middle East, Africa and Asia, geopolitical pressures and northern Europe's relative immunity from the immediate impacts will make the region a prime target for regional and international migration. As declining resource quality exerts a hard brake on Europe's already tepid economic growth, increasing inflows of migrants toward northern Europe will fuel nationalist sentiments and further undermine the territorial integrity of the EU project.

10.2 North America

During this period, US oil production is forecast to increase—but for how long and what extent? US conventional oil production is in decline, meaning that the increasing shortfall is being made-up using unconventional sources (see Fig. 10.5).

Although there has been a large increase in US oil production from 2008 to 2015, from late 2015 to August 2016 that production has faltered, and in fact decreased by 12 % (EIA 2016). A study for the Research Centre for Energy Management for the ESCP Europe Business School finds that although US shale oil production is projected to increase from about 1 mbd in 2012 to 2 mbd in 2020 (and possibly 3 mbd by 2025), this increase would "hardly offset the normal annual depletion rate of 3 %–5 % in US conventional oil production" (Salameh 2012).

The most detailed reality check on the US shale gas situation came from the Post Carbon Institute. A commentary on the PCI's report in *Nature* noted: "Production costs in many shale-gas plays exceed current gas prices, and maintaining production requires ever-increasing drilling and the capital input to support it. Although the extraction of shale gas and tight oil will continue for a long time at some level, production is likely to be below the exuberant forecasts from industry and government" (Hughes 2013).

There remain strong reasons to remain skeptical of hype over US shale gas prospects. US experience suggests that shale gas plays are very heterogeneous, whereas most assessments of recoverable resources assume that productivity is uniform

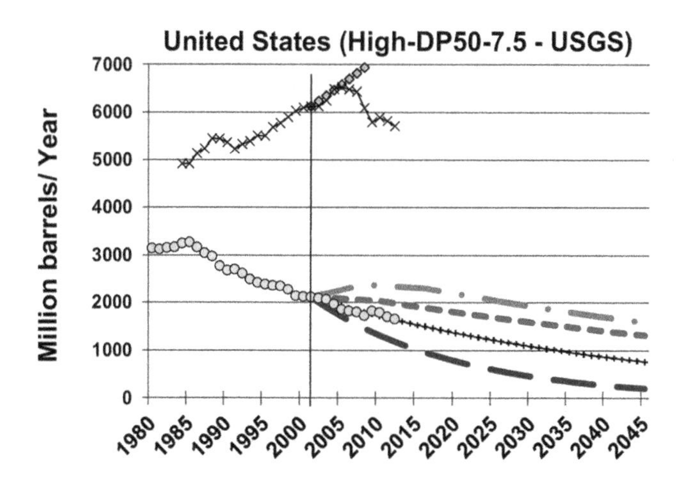

Fig. 10.5 Oil production forecasts for the US. *Source*: Hallock Jr. et al. (2014)

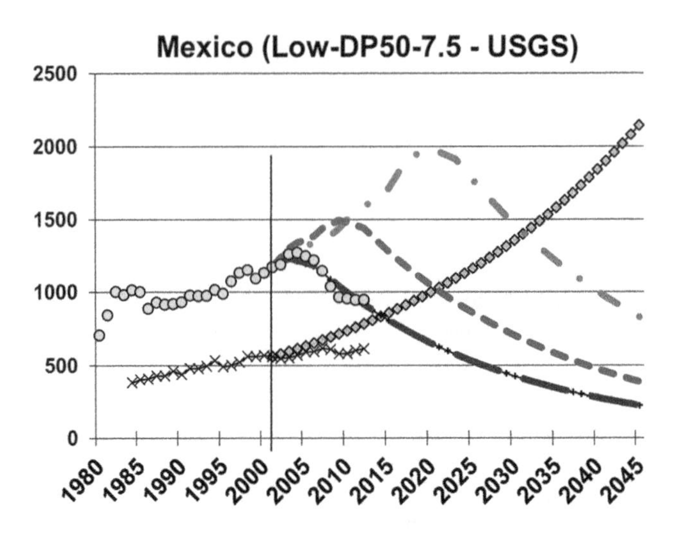

Fig. 10.6 Oil production forecasts for Mexico. *Source*: Hallock Jr. et al. (2014)

across the play, meaning that they "are likely to have significantly overestimated the recoverable resource." Similarly, the use of incorrect functional forms for the 'decline curve' for individual shale gas wells means their recoverable resources could have been overestimated: "Taken together, these considerations suggest that recent US resource estimates may need to be downwardly revised" (McGlade et al. 2013).

When a more specific analysis is undertaken, such as that by the University of Texas, Austin, Department of Petroleum and Geosystems Engineering, US natural gas production from the four biggest plays is far more likely to peak in 2020 (Patzek et al. 2013; Gülen et al. 2013). These studies contradict the exuberant predictions of the US Energy Information Administration (Inman 2014). Renowned oil geologist Arthur Berman, who correctly predicted that the plummeting gas prices (due to initially high production levels at shale gas wells which rapidly taper off requiring continuous drilling at new sites) would render much of the industry unprofitable, states that there are only "years, not decades" left of shale reserves in the US. He forecasts that both shale oil and gas in US will peak in 2025 (Berman and Leonard 2015).

Mexico's oil production peaked in 2006, and is now declining at around 6 % per year (see Fig. 10.6). As its domestic consumption needs have grown, it is increasingly consuming its own oil, which has reduced its export capacity. If that continues, by 2020, Mexico's oil export capacity—most of which is directed toward the US—will become negligible.

This places Mexico on a trajectory toward systemic state-failure between 2020 and 2035. If US-Mexico border issues have been considered problematic to date, they will become truly intractable in that time-frame. In Canada, conventional oil production peaked around 2009, while unconventional production has increased but is likely to reach a plateau in the next few years (also see Fig. 10.7).

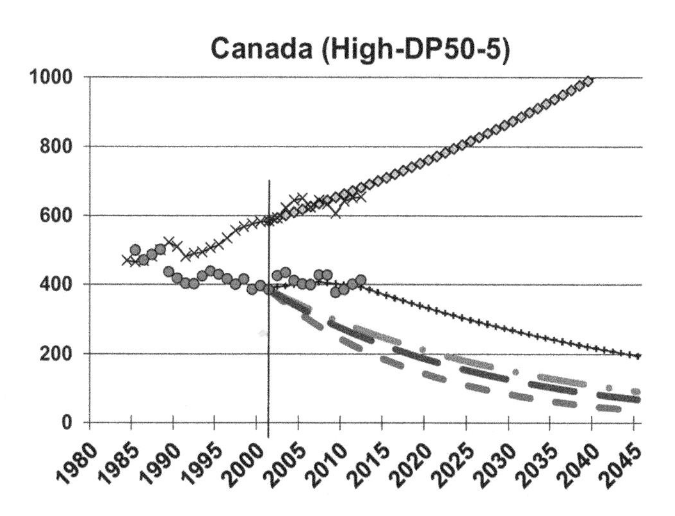

Fig. 10.7 Oil production forecasts for Canada. *Source*: Hallock Jr. et al. (2014)

While the US EIA forecasts that North American production as a whole will increase in the coming decade, an independent model in the *BioPhysical Economics and Resource Quality* journal based on a methodology more closely fitting historical production data concludes that "the combined conventional and unconventional crude oil production from North America is estimated to decline from about 14.7 mbd (2014) to about 13 mbd (2020) and 10 mbd (2030)" (Dittmar 2016).

Unlike Northern Europe, North America will be quite vulnerable to converging climate impacts in terms of water scarcity and reductions in crop yields. In coming decades, the Southwest and Central Plains will face increased drought severity, which will intensify in the later half of the twenty-first century (Cook et al. 2015). Climate change is also increasing the frequency, magnitude, and duration of drought in California, the US' major food basket. The record drought has involved acute water shortages, enhanced wildfire risk, and is endangering agriculture. A study in the *Proceedings for the National Academy of Sciences* concludes that "anthropogenic warming is increasing the probability of co-occurring warm–dry conditions like those that have created the acute human and ecosystem impacts associated with the 'exceptional' 2012–2014 drought in California" (Diffenbaugh et al. 2015).

The California drought has so far produced the greatest water loss ever seen in California agriculture, reducing river water for Central Valley farms by a third. Groundwater is being increasingly depleted to make up for the water loss in agricultural and other industrial uses: "Groundwater is being pumped at far greater rates than it can be naturally replenished, so that many of the largest aquifers on most continents are being mined, their precious contents never to be returned" (Famiglietti 2014). Lake Mead, the largest reservoir in the US which supplies Los Angeles and 90 % of Southern Nevada, has now been depleted to just 37 % of its former size. (Pierce 2016) There is a 50 % chance that climate change and overuse will cause Lake Mead to dry up by 2021. By 2050 the Colorado River is likely to experience a

10–30 % drop in the amount of runoff it receives from snow that falls and melts on the western slope of the Rocky Mountains. This would also lead to its upstream twin, Lake Powell, becoming permanently dry (Barnett and Pierce 2008).

This rate of groundwater depletion threatens the viability of California's agriculture, which would hit national supplies of grains, fruit and vegetables within the US, as well as US exports. It is also making an increasingly large dent in California state revenues ($2.7 billion as of 2015), and as groundwater dries up to a point-of-no-return, this will force up local and national water and food prices as California is compelled to begin purchasing expensive water and food imports nationally and from abroad. That means by 2025, the US will begin to experience a deepening national water and food crisis. This will also create a permanent dent in the global food system, effectively removing a large chunk of US food exports from world food markets, thus contributing to a resurgence of global food prices that would likely exceed the unprecedented spikes of 2011. Food prices could rise as high as 395 % after 2020 (Ahmed 2016b).

Yet this will be compounded by a creeping national energy crisis as the US' chief domestic unconventional oil and gas sources lose their ability to compensate for the shortfall in conventional production. The US will once again be forced to rely on increasing energy imports, yet as we approach 2030, the world oil market will face a situation of plateauing exports from OPEC and the Middle East even as demand rises dramatically in Asia.

10.3 Post-2030–2045

Faced with these converging crises, the Euro-Atlantic core will continue to see the creation of cheap debt-money through quantitative easing as an immediate solution to generate emergency funds to stabilize the financial system and shore-up ailing industries. This would likely play out in either of two potential business-as-usual scenarios:

1. The lower resource quality (EROI) of the global energy system may act as a fundamental geophysical ceiling on the capacity of the economy to grow. It may act as an invisible brake on growth in demand, potentially meaning that fossil fuel prices would remain at chronically low levels endangering the profitability of the fossil fuel industries. This would lead to an acceleration of the demise of the fossil fuel industries, which would in itself raise the question of debt-defaults across the industries feeding back into the financial system. Endangered hydrocarbon energy production would feed back into the global system as a further lowering of resource quality, manifesting in a protracted and self-reinforcing recessionary economic process. This would escalate vulnerability to water, food and energy crises and hugely strain the capacity of European and American states to deliver goods and services to their own populations.

2. Scarcity of net exports on the world market may manifest in a resurgence of oil prices. This may return some sectors of ailing fossil fuel industries to a modicum of profitability. However, previous slashing of investments and cutbacks in exploration—much of which has already taken place—will mean that only the most powerful sections of the industry would be able to capitalize on this, meaning that production is unlikely to be able to return to previously high levels. Price spikes would trigger economic recession, causing a drop in demand, while lower production levels would exacerbate the economy's inability to grow substantially, if at all. In effect, the global economy would likely still experience a self-reinforcing recessionary economic process.

In both scenarios, escalating economic crises are likely to invite the Euro-Atlantic core to respond by using debt-money to shore-up as much of the existing core financial and energy industries as possible. Prices spikes and shortages in water, food and energy would be experienced by general populations as a dramatic lowering of purchasing power, leading to an overall decrease in quality of life, an increase in poverty, and a heightening of inequality. This would undermine their internal cohesion, giving rise to new divisive, nationalist and xenophobic movements, and lead states into a tightening spiral of militarization to police domestic order. As instability in the Middle East and elsewhere intensifies, manifesting in further unrest, political violence and terrorist activity, states will also be drawn increasingly into short-sighted military solutions. In particular, scarcity of net oil exports on the world market will heighten geopolitical and military competition to control and/or access the world's remaining hydrocarbon energy resources. With the Middle East still holding the vast bulk of the world's reserves, the region will remain a central flashpoint for such competition, even as major producers such as Saudi Arabia approach systemic state-failure due to reaching inevitable production declines.

Both these scenarios could be dramatically ameliorated with productive investments of remaining financial resources in new renewable energy systems, along with more sustainable forms of water management and small-scale organic agriculture—combined with fundamental political economic restructuring—instead of reactionary efforts to save the institutions at the heart of responsibility for the crises (Ahmed 2010). In the absence of such investments and re-structuring, it is difficult to avoid the conclusion that as we near 2045, the European and American projects will face escalating internal challenges to their internal territorial integrity, increasing the risk of systemic state-failure.

Chapter 11
Conclusions: From Systemic State-Failure to Civilizational Transition

11.1 Global Phase-Shift

The cases of crisis convergence already occurring and examined here provide concrete empirical evidence on the impacts of Earth System Disruption (ESD) in terms of the breakdown of human societies, or Human System Destabilization (HSD). While much of the West has so far avoided the most important aspects of such disruptions the writing is on the wall that no country can feel completely safe from its potential impacts. Rather than ESD eventually unfolding in some far-flung, catastrophic imaginary future to be pondered purely through abstract models, it is destabilizing the local state-level sub-systems of key regions here and now.

Each case study explored here represents the reality of emerging ESD–HSD along with local amplifying feedback processes, where the convergence of energy and environmental crises feeds into, and in turn is fueled by, political and economic crises. Poorer, less developed countries on the periphery of the world system find themselves more politically and economically weaker, and thus more vulnerable to ESD processes which happen to be more pronounced in these regions. As such, biophysical ESD processes have triggered interlinked social, political and economic crises leading to the breakdown of traditional order (HSD). In turn, HSD rapidly breaks down the capacity of local states to respond effectively to the ESD processes triggering HSD in the first place, allowing ESD to continue accelerating. The acceleration of ESD, in turn, establishes a biophysical groundwork for the continued triggering of HSD. This overarching process, in effect, amounts to a local ESD–HSD amplifying feedback loop, in which the intertwined nature of ESD and HSD processes becomes self-reinforcing. Within the last decade, local ESD–HSD amplifying feedback loops have broken out in major countries in every key continent.

Taking a holistic, planetary perspective suggests that local ESD–HSD amplifying feedbacks have already reached a global pandemic scale, even though this pandemic still remains in a nascent state. As crisis convergence accelerates, these ESD–HSD processes, which are already interconnected across the global system in

N.M. Ahmed, *Failing States, Collapsing Systems*, SpringerBriefs in Energy,
DOI 10.1007/978-3-319-47816-6_11

complex ways, will intensify and interlock further. They are, ultimately, symptoms of a wider systemic process: an accelerating phase-shift toward a new systemic configuration that is being driven by the global system's increasing collision with its own biophysical basis.

The intensification of political violence can be seen as a symptom and direct result of this overarching global systemic crisis. The rise in Western militarism as well as Islamist militancy are 'securitized' HSD responses to ESD processes that are unravelling the very basis of order as we know it. To be more precise, the intensification of HSD appears, in effect, to be a direct result of the thermodynamics of accelerating energy consumption, depletion and dissipation—which is trending rapidly toward a threshold effect.

The time-scale to reach this global system threshold effect cannot be pinned down, but an important finding of this study is that given the hindsight provided by the cases in the Middle East, it takes about 15 years before a government experiences systemic state-failure from the point at which its energy and economic base begins to decline, relative to climate-induced water and food crises. This in turn suggests that after 2030, both the Euro-Atlantic core, as well as the fast-rising Indo-Chinese periphery, will begin to experience their own symptoms of systemic state-failure.

The cases examined here thus point to a global process of civilizational transition. As a complex adaptive system, human civilization in the twenty-first century finds itself at the early stages of a systemic phase-shift which is already manifesting in local sub-system failures in every major region of the periphery of the global system. As these sub-system failures driven by local ESD–HSD amplifying feedbacks accelerate and converge in turn, they will coalesce and transmit ever more powerfully to the core of the global system. As this occurs and re-occurs, it will reach a system-wide threshold effect resulting in eventual maladaptive global system failure; or it will compel an adaptive response in the form of fundamental systemic transformation.

In this context, it is precisely the acceleration of global system failure that paves the way for the possibility of fundamental systemic transformation, and the emergence of a new phase-shift in the global system. Far from representing the end of the human species, these processes represent the inevitable demise of a historically-specific global systemic configuration in the form of neoliberal finance capitalism as it transitions into a new era dominated by information technologies. Well before 2050, this study suggests, systemic state-failure will have given way to the irreversible demise of neoliberal finance capitalism as we know it.

Human civilization is in the midst of a global transition to a completely new system which is being forged from the ashes of the old. Yet the contours of this new system remain very much subject to our choices today. If the forces of systemic failure overwhelm us, then the new systemic configuration is likely to represent a maladaptive collapse in civilizational complexity. Yet even within such a maladaptive response—which arguably is well-underway as these cases show—there remains a capacity for agents within the global system to generate adaptive responses that, through the power of transnational information flows, hold the potential to enhance collective consciousness. The very breakdown of the prevailing system heralds the potential for long-term post-breakdown systemic transformation.

11.2 Clean Energy and Environmental Restoration

Due to the centrality of fossil fuels to the dysfunctional nature of our global systemic predicament, it is obvious that an integral component of a successful transition to a new, healthier and more sustainable civilizational phase-shift must involve a concerted shift away from fossil fuels. This requires on the one hand an accelerating investment in cleaner, alternative energy technologies that are renewable. But it also requires a reduction in energy consumption that recognizes that the neoclassical ideology of endless growth that animates conventional economic thinking must be rejected.

That further requires a level of environmental consciousness that looks to the long-term in a serious way. It is quite possible that a worst-case business-as-usual scenario for climate change, for instance, does indeed transpire. Yet this is no reason to resort to a fatalistic response. On the contrary, it requires pushing forward the vision for systemic transformation of the human species many centuries into the future, to a point when the planet is able to return to some form of equilibrium. Whatever that future equilibrium looks like, whatever levels of existence are possible in such a context, human societies will require principles and modes of organization very different to what is taken for granted today. The thinking and technologies for that future will still need to be planted.

In the meantime, the evidence discussed here demonstrates that substantive renewable energy investments are not merely a societal and economic dividend, but constitute one of the most important long-term sources of resilience from the destabilizing potential of system failure. Yet renewables are not a panacea. They do not provide the potential for unlimited material production and consumption that is integral to capitalism as we know it, and to transition to them rapidly still requires vast levels of investment for which the political will does not presently exist. However, as production costs of renewables continue to drop even as their technological efficiency increases, so will the potential for grassroots communities to rally around such technologies. To that extent, they may also help catalyze wider necessary political and economic structural transformations.

In short, this analysis suggests that rapid sustainable development of a global 100 % renewable energy system is among the most urgent issues for policymakers this century. The technical and logistical potential for this exists, even if it is incommensurate with an endless growth economic model. In fact, such a transition is possible within 20–30 years using less than 40 % of proven reserves of conventional oil (Schwartzman and Schwartzman 2013). Simultaneously, and equally urgent, is a rapid shift toward forms of food and water production and distribution which are completely independent from fossil fuels. Low intensity organic agriculture, which is less water intensive and free of fertilizers and pesticides, can potentially be scaled up to provide sufficient calories to feed the whole human population as it eats today (Badgley and Perfecto 2007). Researchers at Washington State University have more recently found that a mix of non-conventional agricultural methods—organic, agroforestry, integrated farming, conservation agriculture, mixed crop/livestock, among others—could feed the world, weaning the global food system off fossil fuel dependence, and producing food even in severe drought conditions. There is also a dire need to reduce food waste—as much as 30–40 % of all food produced globally is wasted, demonstrating the potential for wasted food in over consuming nations to be redistributed for consumption by poorer and hungrier populations (Reganold and Wachter 2016).

11.3 Circular Economy and Post-Capitalism

Concepts of the 'circular economy'—involving a fundamental reorganization of the way societies produce, manage and consume resources through wide-scale practices of recycling across production and consumption chains—bear considerable importance to this sort of vision. The circular economy brings to the fore the necessity of reusing and recycling raw materials to the most efficient extent possible to support the sustainability of production and consumption chains relative to increasingly depleted mineral ores and higher energy costs for their extraction, refining and input into manufacturing. Numerous companies are taking the concept seriously in the recognition of current and looming environmental risks to their supply chains, but human civilization must begin to do so in the wider context of a recognition that the animating ideology of the current phase-shift of civilization is deeply misguided.

A major report to the Club of Rome tracking the depletion of the planet's mineral ores finds that by the end of this century, higher quality ores critical to the growth of industrial civilization as we know it will be largely depleted. But the report shows that a 'circular economy' approach has strong potential to allow existing minerals to be recycled with minimum losses and a high degree of efficiency sufficient to maintain a high technology society in, however, a new post-capitalist economic context (Bardi 2014).

The ESD–HSD amplifying feedbacks discussed in this study demonstrate that the twenty-first century is rapidly transitioning to a crisis convergence threshold heralding the inevitable demise of the endless growth model of neoliberal finance capitalism that currently animates industrial civilization as we know it. This points to the urgency of adaptation to prepare for the emergence of a new evolutionary phase-shift in the form of post-capitalism—a concept whose unspecified nature is important precisely because it opens up new possibilities for economic organization which are not limited by the failures of prevailing economic orthodoxy. The rejection of that orthodoxy as limited springs from the recognition that the doctrine of unlimited economic growth is nothing less than a fundamental violation of the laws of physics. In short, it is the stuff of cranks—yet it is nevertheless the ideology that informs policymakers and pundits alike. Post-capitalism, on the other hand, seeks to ground itself in harmony with the biophysical environment, not by rejecting the ideals of human prosperity and well-being, but by decoupling them from the fetish for endless material growth.

That in turn paves the way, potentially, for a renewed sense of human value and purpose beyond the confines of material production and consumption, rejuvenated by a consciousness of humanity's embeddedness in its environment. The magical thinking of endless growth must make way for a post-materialist ethic of human interconnectedness with itself and its biophysical context.

Groundbreaking economic work on this theme of 'prosperity without growth' has been forthcoming from several quarters, and provides mounting evidence that the endless growth model of economics has not just failed to deliver meaningful prosperity to the world's poorest, but is incapable of doing so as it continues to generate inequality, environmental destruction, and to eventually undermine its very own basis (Jackson 2009). This body of work also demonstrates that meaningful prosperity in the sense of providing for human needs and well-being in high technology societies remains possible in a fundamentally re-organized post-capitalist economy. In this framework, human progress can continue but within a new paradigm in which limited material development is mobilized to meet fundamental human needs through extension of human relations instead of market relations, a deepening of democracy, enhancing ecosystems, and more equal distribution of wealth. Inevitably, therefore, post-capitalism will be incommensurate with the features of endless growth associated with industrial forms of capitalism: namely, continuously growing material throughput driven by ever growing consumption by unrestrained population growth (Victor 2010; Fournier 2008; Schneider et al. 2010; Fritz and Koch 2014).

Instead, the unsustainable nature of contemporary capitalism opens up the urgency of working toward a new post-capitalist era built on the following components: regulation of market mechanisms and corporate activities; support for social enterprises organized as community cooperatives; democratic money creation processes, including community currencies, in place of debt-based fractional reserve banking; communities reclaiming the commons, especially in the sense of communal land stewardship systems; redistribution of income and capital assets; a diversity of production scales and modes, including small-scale, subsistence and self-employment to widen economic democracy (Johanisova and Wolf 2012).

Such a vision may, in the current context, appear impossibly utopian. By 2030, and even more so by 2050—as the manifestations of global capitalism's self-catabolic trajectory become more obvious—it will appear increasingly realistic.

11.4 Information Revolution and Social Liberation

The political will for an emergency effort to effect such a transition to post-carbon, post-capitalist civilizational forms does not yet exist. Part of the reason for this is the whole systems knowledge deficit. Despite an abundance of information, there is a paucity of actionable knowledge which translates this information into a holistic understanding of the nature of the current global phase-shift and its terminal crisis trajectory for all relevant stakeholders. While much of the human population has been denied access to such information, and thus actionable knowledge, vested interests in the global fossil fuel and agribusiness system are actively attempting to control information flows to continue to deny full understanding in order to perpetuate their own power and privilege. The only conceivable pathway out of this impasse, however difficult or unlikely it may appear, is to break the stranglehold of information control by disseminating knowledge on both the causes and potential solutions to global crisis.

In the absence of accurate interpretations of the information we have—actionable knowledge—human civilization collectively is bound to pursue a dysfunctional maladaptive path toward protracted global system failure. Yet the current phase-shift of neoliberal finance capitalism has also, for the first time in human history, generated an opportunity through the information technology revolution for vast numbers of people from all over the world to become connected and informed through the internet. The impacts of this are already palpable, in the sense that larger numbers of educated citizens than ever before around the world are aware of the risks of various global crises from climate change, to economic inequality. However, there is very little in the way of a holistic understanding of the biophysical triggers for these crises, their fundamental interconnections, as well as their aggravating causes in the present structure of human civilization—a matter in which the vast majority of global population remains in deep denial.

Meanwhile, the opportunity of the internet is being rapidly coopted by the Global Media-Industrial Complex. Despite this, the inherently decentralized dynamic of the internet means that with every effort of top-down cooptation, the potential for bottom-up innovation increases. This means that perhaps the most important imperative is to use the information revolution to intervene in and compete with the inaccurate information flows of the GMIC, by generating new more accurate networks of communication based on transdisciplinary knowledge which is, most importantly, translated into user-friendly multimedia information widely disseminated and accessible by the general public in every continent. This actionable knowledge must, in holistic fashion, encompass both the diagnosis of the global problem, as well as the prognosis in the form of exploratory visions and practical actions for active participation in civilizational transition to an inevitable post-carbon future.

The systemic target for such counter-information dissemination, moreover, is eminently achievable. Social science research has demonstrated that the tipping point for minority opinions to become mainstream, majority opinion is 10 % of a given population (Phys.org 2016). The challenge in the context of global system change is that this requires not simply 10 % of a given population being actionably convinced of only one component of the crisis—such as climate change: that 10 % must be convinced of the reality and solution to all of these components, simultaneously, and systemically.

This is, of course, a major challenge, but underscores that any meaningful strategy to generate widespread social action in support of a phase-shift toward systemic transformation must reach 10 % of any given population, and be grounded in holistic-systemic diagnosis and prognosis.

By coupling an information strategy targeting 10 % of a given population with concrete actions to generate forms of change that are meaningful for individuals and communities, tackling the knowledge deficit can simultaneously contribute to establishing the local groundwork for an adaptive systemic response. This can help build local resilience to local sub-system failures, while also pushing back against the growing global ESD–HSD pandemic, and potentially even facilitating the pandemic to pave the way for renewed adaptive responses.

11.5 A New Action-Research Agenda

These findings establish the urgency of a transdisciplinary action-research agenda across the natural and social sciences. Such a new program must focus on excavating and integrating the dynamics of sub-system failure across multiple domains from climate, to energy, to economics, to explore the tangible, real-world actualities of how such processes are playing out today.

However, this call for a new action-research agenda comes with caveats. This study highlights the pitfalls of predictive modelling as the prime focus of scientific research on global crises and their potential to generate political violence. For the most part, not a single one of the cases discussed here was ever modelled or predicted by any scientist or scholar. Due to the sheer complexity of the phenomena under examination, it is unlikely that any single model alone would be capable of realistically capturing all their features in a way that would produce a meaningful model-run of their crisis-trajectories. This is particularly important to understand because of the one critical factor that no model can fully capture—human agency. A full and accurate model of the scope for human agency's input into various crisis-trajectories would require a vast amount of historical, sociological, geopolitical, economic and cultural data, most of which could never be quantified. It is for this reason that a renewed effort must be made for modelers and theoreticians to come together in integrating their work to create more robust paradigms by which to understand human societies.

The most obvious port of call for funding for this new action-research agenda includes national governments, local authorities, and intergovernmental institutions, many of which have departments with some interests in funding such issues. Military, intelligence and security agencies with a direct interest in developing intelligence to assess contemporary challenges and forecast future events and scenarios should be at the forefront of such funding in order to augment and rapidly rectify their existing analytical capabilities. Given the scale and complexity of the issues, these crises have already overwhelmed their abilities to make sense of them enough to properly inform viable national and international security policies. Emerging post-carbon, post-capitalist industries, such as the renewable energy sector, may also be quick to recognize the importance of such an action-research program.

This study has focused on elaborating an empirically-formed theoretical framework for such research, which suggests several avenues for further work and scholarly-practitioner collaboration. My focus here has been on demonstrating the biophysical basis of contemporary geopolitical chaos in every major region of the world. This is not to reduce geopolitical conflict purely to biophysical triggers, but to recognize that the geopolitical—like the social and economic—remains indelibly embedded in the biophysical. The failure to understand this, in turn, remains chiefly responsible for the impotence with which policymakers approach accelerating geopolitical crises. Further research is required to draw on multiple political, social science and economic theories to develop the theoretical contours of this framework. Equally, we have identified a theoretical basis for more integrated transdisciplinary statistical modelling that integrates data across traditionally separate domains. However, neither of these avenues can succeed if they proceed separately. For models to be useful and generate real predictive and/or explanatory value, they must be grounded in rigorous conceptual and theoretical architectures that are reflective of the real world. And for theories to actually reflect the real world, they must draw on valid empirical data.

This new research must also be tied to practical actions aimed at generating system change: whether that means creating new cultures of collaboration to disseminate research to a wider audience, or tying scholarship to real world initiatives involving communities engaged in various forms of system change, or seeking to influence policy by engaging with relevant institutions—research must be action-oriented.

Bibliography

Adel, Mohamed. 2016. Eni to Increase Zohr Field Gas Production to 2bn Cubic Feet Per Day by End of 2019. *Daily News Egypt*, May 9. http://www.dailynewsegypt.com/2016/05/09/eni-increase-zohr-field-gas-production-2bn-cubic-feet-per-day-end-2019/.

Agrimoney. 2012. Unrest, Bad Weather Lift Syrian Grain Import Needs. *Agrimoney.com*, March 14. http://www.agrimoney.com/news/unrest-bad-weather-lift-syrian-grain-import-needs--4278.html.

Ahmed, Nafeez Mosaddeq. 2009. The Globalization of Insecurity: How the International Economic Order Undermines Human and National Security on a World Scale. *Historia Actual Online* 0(5): 113–126.

Ahmed, Nafeez. 2010. *A User's Guide to the Crisis of Civilisation: And How to Save It*. London: Pluto Press.

———. 2011. The International Relations of Crisis and the Crisis of International Relations: From the Securitisation of Scarcity to the Militarisation of Society. *Global Change, Peace & Security* 23(3): 335–355. doi:10.1080/14781158.2011.601854.

———. 2013a. Peak Oil, Climate Change and Pipeline Geopolitics Driving Syria Conflict. *The Guardian*, May 13, sec. Environment. https://www.theguardian.com/environment/earth-insight/2013/may/13/1.

———. 2013b. How Resource Shortages Sparked Egypt's Months-Long Crisis. *The Atlantic*, August 19. http://www.theatlantic.com/international/archive/2013/08/how-resource-shortages-sparked-egypts-months-long-crisis/278802/.

———. 2014. Behind the Rise of Boko Haram—Ecological Disaster, Oil Crisis, Spy Games. *The Guardian*, May 9, sec. Environment. https://www.theguardian.com/environment/earth-insight/2014/may/09/behind-rise-nigeria-boko-haram-climate-disaster-peak-oil-depletion.

———. 2015. The US-Saudi War with OPEC to Prolong Oil's Dying Empire. *Middle East Eye*. May 8. http://www.middleeasteye.net/columns/us-saudi-war-opec-prolong-oil-s-dying-empire-222413845.

———. 2016a. Climate Change Fuels Boko Haram. *Women Across Frontiers Magazine*. February 29. http://wafmag.org/2016/02/boko-haram-filling-vacuum-nigerias-state-collapses/.

———. 2016b. At the Root of Egyptian Rage Is a Deepening Resource Crisis. *Quartz*. Accessed August 16. http://qz.com/116276/at-the-root-of-egyptian-rage-is-a-deepening-resource-crisis/.

———. 2016c. Return of the Reich: Mapping the Global Resurgence of Far Right Power. Investigative Report. London: Tell MAMA and INSURGE Intelligence. https://medium.com/return-of-the-reich.

N.M. Ahmed, *Failing States, Collapsing Systems*, SpringerBriefs in Energy,
DOI 10.1007/978-3-319-47816-6

———. 2016d. FEMA Contractor Predicts 'Social Unrest' Caused by 395% Food Price Spikes. *Motherboard*. Accessed August 21. http://motherboard.vice.com/read/fema-contractor-predicts-social-unrest-caused-by-395-food-price-spikes.

Akuru, Udochukwu B., and Ogbonnaya I. Okoro. 2011. A Prediction on Nigeria's Oil Depletion Based on Hubbert's Model and the Need for Renewable Energy. *International Scholarly Research Notices, International Scholarly Research Notices* 2011: e285649. doi:10.5402/2011/285649.

Al-Sinousi, Mahasin, and Amira Saleh. 2008. International Expert Warns Of Egypt's Oil And Gas Reserves Depletion In 2020. *Al-Masry Al-Youm*, May 17, 1434 edition. http://today.almasry-alyoum.com/article2.aspx?ArticleID=105585.

Arashi, Fakhri. 2013. Wheat Imports Cause Yemen Heavy Losses—National Yemen. http://nation-alyemen.com/2013/03/03/wheat-imports-cause-yemen-heavy-losses/.

Aston, T.H., Trevor Henry Aston, and C.H.E. Philpin. 1987. *The Brenner Debate: Agrarian Class Structure and Economic Development in Pre-Industrial Europe*. Cambridge: Cambridge University Press.

Aucott, Michael L., and Jacqueline M. Melillo. 2013. A Preliminary Energy Return on Investment Analysis of Natural Gas from the Marcellus Shale. *Journal of Industrial Ecology* 17(5): 668–679. doi:10.1111/jiec.12040.

Azevedo, Ligia B., An M. De Schryver, A. Jan Hendriks, and Mark A.J. Huijbregts. 2015. Calcifying Species Sensitivity Distributions for Ocean Acidification. *Environmental Science & Technology* 49(3): 1495–1500. doi:10.1021/es505485m.

Badgley, Catherine, and Ivette Perfecto. 2007. Can Organic Agriculture Feed the World? *Renewable Agriculture and Food Systems* 22(2): 80–85.

Bardi, Ugo. 2014. *Extracted: How the Quest for Mineral Wealth Is Plundering the Planet*. Vermont: Chelsea Green Publishing.

Barnett, Tim P., and David W. Pierce. 2008. When Will Lake Mead Go Dry? *Water Resources Research* 44(3): W03201. doi:10.1029/2007WR006704.

Barron, Robert. 2016. Facing Rumors of Money Troubles, Egypt Denies Tension with Foreign Oil, Gas Firms. *Mada Masr*. January 27. http://www.madamasr.com/sections/economy/facing-rumors-money-troubles-egypt-denies-tension-foreign-oil-gas-firms.

Berger, Daniel, William Easterly, Nathan Nunn, and Shanker Satyanath. 2013. Commercial Imperialism? Political Influence and Trade during the Cold War. *American Economic Review* 103(2): 863–896. doi:10.1257/aer.103.2.863.

Berman, Arthur, and Ray Leonard. 2015. Years Not Decades: Proven Reserves and the Shale Revolution. *Houston Geological Society Bulletin* 57(6): 35–39.

Bhardwaj, Mayank. 2016. Food Imports Rise as Modi Struggles to Revive Rural India. *Reuters India*. February 2. http://in.reuters.com/article/india-farming-idINKCN0VA3NL.

Bindi, Marco, and Jørgen E. Olesen. 2010. The Responses of Agriculture in Europe to Climate Change. *Regional Environmental Change* 11(1): 151–158. doi:10.1007/s10113-010-0173-x.

Bose, Prasenjit. 2016. A Budget That Reveals the Truth about India's Growth Story. *The Wire*. March 2. http://thewire.in/23392/what-the-budget-tells-us-about-indias-growth-story/.

Boucek, Christopher. 2009. Yemen: Avoiding a Downward Spiral. *Carnegie Endowment for International Peace*. September. http://carnegieendowment.org/2009/09/10/yemen-avoiding-downward-spiral-pub-23827.

Bove, Vincenzo, Leandro Elia, and Petros G. Sekeris. 2014. US Security Strategy and the Gains from Bilateral Trade. *Review of International Economics* 22(5): 863–885. doi:10.1111/roie.12141.

Bove, Vincenzo, Kristian Skrede Gleditsch, and Petros G. Sekeris. 2015. 'Oil above Water' Economic Interdependence and Third-Party Intervention. *Journal of Conflict Resolution, January* 27: 0022002714567952. doi:10.1177/0022002714567952.

Bove, Vincenzo, and Petros G. Sekeris. 2016. Fueling Conflict: The Role of Oil in Foreign Interventions. *IPI Global Observatory*. Accessed July 19. https://theglobalobservatory.org/2015/03/civil-wars-oil-above-water-military-intervention/.

Brandt, Adam R., Yuchi Sun, Sharad Bharadwaj, David Livingston, Eugene Tan, and Deborah Gordon. 2015. Energy Return on Investment (EROI) for Forty Global Oilfields Using a Detailed Engineering-Based Model of Oil Production. *PLOS ONE* 10(12): e0144141. doi:10.1371/journal.pone.0144141.

Brown, Jeffrey J., and Samuel Foucher. 2008. A Quantitative Assessment of Future Net Oil Exports by the Top Five Net Oil Exporters. *Energy Bulletin*. January 8. http://www.resilience.org/stories/2008-01-08/quantitative-assessment-future-net-oil-exports-top-five-net-oil-exporters.

Brown, James H., William R. Burnside, Ana D. Davidson, John P. DeLong, William C. Dunn, Marcus J. Hamilton, Norman Mercado-Silva, et al. 2011. Energetic Limits to Economic Growth. *BioScience* 61(1): 19–26. doi:10.1525/bio.2011.61.1.7.

Buckley. 2016. Coal Decline Steepens in 2016 in India, China, U.S. *Institute for Energy Economics & Financial Analysis*. May 16. http://ieefa.org/coal-decline-steepens-2016-2/.

Capellán-Pérez, Iñigo, Margarita Mediavilla, Carlos de Castro, Óscar Carpintero, and Luis Javier Miguel. 2014. Fossil Fuel Depletion and Socio-Economic Scenarios: An Integrated Approach. *Energy* 77: 641–666. doi:10.1016/j.energy.2014.09.063.

Castillo-Mussot, Marcelo del, Pablo Ugalde-Véle, Jorge Antonio Montemayor-Aldrete, Alfredo de la Lama-García, and Fidel Cruz. 2016. Impact of Global Energy Resources Based on Energy Return on Their Investment (EROI) Parameters. *Perspectives on Global Development and Technology* 15(1–2): 290–299. doi:10.1163/15691497-12341389.

Chen, Shuai, Xiaoguang Chen, and Xu. Jintao. 2016. Impacts of Climate Change on Agriculture: Evidence from China. *Journal of Environmental Economics and Management* 76: 105–124. doi:10.1016/j.jeem.2015.01.005.

Chowdhury, Shakhawat, and Muhammad Al-Zahrani. 2013. Implications of Climate Change on Water Resources in Saudi Arabia. *Arabian Journal for Science and Engineering* 38(8): 1959–1971. doi:10.1007/s13369-013-0565-6.

Clarkson, M.O., S.A. Kasemann, R.A. Wood, T.M. Lenton, S.J. Daines, S. Richoz, F. Ohnemueller, A. Meixner, S.W. Poulton, and E.T. Tipper. 2015. Ocean Acidification and the Permo-Triassic Mass Extinction. *Science* 348(6231): 229–232. doi:10.1126/science.aaa0193.

Cleveland, Cutler J., and Peter A. O'Connor. 2011. Energy Return on Investment (EROI) of Oil Shale. *Sustainability* 3(11): 2307–2322. doi:10.3390/su3112307.

Coleman, Isabel. 2012. Reforming Egypt's Untenable Subsidies. *Council on Foreign Relations*. April 6. http://www.cfr.org/egypt/reforming-egypts-untenable-subsidies/p27885.

Cook, Benjamin I., Toby R. Ault, and Jason E. Smerdon. 2015. Unprecedented 21st Century Drought Risk in the American Southwest and Central Plains. *Science Advances* 1(1): e1400082. doi:10.1126/sciadv.1400082.

Coumou, Dim, Alexander Robinson, Stefan Rahmstorf. 2013. Global increases in record-breaking monthly-mean temperatures. Climatic Change 118(3): 771-782. doi:10.1007/s10584-012-0668-1.

Csereklyei, Zsuzsanna, and David I. Stern. 2015. Global Energy Use: Decoupling or Convergence? *Energy Economics* 51: 633–641. doi:10.1016/j.eneco.2015.08.029.

Cunningham, Nick. 2016. Decline of Coal Demand Is 'irreversible. *MINING.com*. February 19. http://www.mining.com/web/decline-of-coal-demand-is-irreversible/.

Dawson, Terence P., Anita H. Perryman, and Tom M. Osborne. 2014. Modelling Impacts of Climate Change on Global Food Security. *Climatic Change* 134(3): 429–440. doi:10.1007/s10584-014-1277-y.

Daya, Ayesha, and Dana El Baltaji. 2016. Saudi Arabia May Become Oil Importer by 2030, Citigroup Says. *Bloomberg.com*. Accessed August 11. http://www.bloomberg.com/news/articles/2012-09-04/saudi-arabia-may-become-oil-importer-by-2030-citigroup-says-1-.

DCDC. 2013. Regional Survey—South Asia Out to 2040. Strategic Trends Programme. UK Ministry of Defence, Defence Concepts and Doctrines Centre.

Department Of State, Bureau of Public Affairs. 2014. Syria. Press Release|Fact Sheet. *U.S. Department of State*. March 20. http://www.state.gov/r/pa/ei/bgn/3580.htm.

Diffenbaugh, Noah S., Daniel L. Swain, and Danielle Touma. 2015. Anthropogenic Warming Has Increased Drought Risk in California. *Proceedings of the National Academy of Sciences* 112(13): 3931–3936. doi:10.1073/pnas.1422385112.

Dipaola, Anthony. 2016. Iraq's Oil Output Seen by Lukoil at Peak as Government Cuts Back. *Bloomberg.com.* May 19. http://www.bloomberg.com/news/articles/2016-05-19/iraq-s-oil-output-seen-by-lukoil-at-peak-as-government-cuts-back.

Dittmar, Michael. 2016. Regional Oil Extraction and Consumption: A Simple Production Model for the Next 35 Years Part I. *BioPhysical Economics and Resource Quality* 1(1): 7. doi:10.1007/s41247-016-0007-7.

Dodge, Robert. 2016. Unconventional Drilling for Natural Gas in Europe. In *The Global Impact of Unconventional Shale Gas Development*, ed. Yongsheng Wang and William E. Hefley, 97–130. Natural Resource Management and Policy 39. Springer International Publishing. http://link.springer.com/chapter/10.1007/978-3-319-31680-2_5.

EASAC. 2014. Shale Gas Extraction: Issues of Particular Relevance to the European Union. European Academies Science Advisory Council. http://www.easac.eu/fileadmin/Reports/EASAC_ExecSummary___Statement_ShaleGas_Extraction_combined.pdf.

Ebrahimi, Mohsen, and Nahid Ghasabani. 2015. Forecasting OPEC Crude Oil Production Using a Variant Multicyclic Hubbert Model. *Journal of Petroleum Science and Engineering* 133: 818–823. doi:10.1016/j.petrol.2015.04.010.

EI. 2012. Youth Are Quarter of Egypt's Population, and Half of Them Are Poor | Egypt Independent. *Egypt Independent*. August 12. http://www.egyptindependent.com/news/youth-are-quarter-egypt-s-population-and-half-them-are-poor.

EIA. 2016. Petroleum & Other Liquids Weekly Supply Estimates. US Energy Information Administration. http://www.eia.gov/dnav/pet/pet_sum_sndw_dcus_nus_w.htm.

Evans-Pritchard, Ambrose. 2015. Saudi Arabia May Go Broke before the US Oil Industry Buckles. *The Telegraph*, August 5, sec. 2016. http://www.telegraph.co.uk/business/2016/02/11/saudi-arabia-may-go-broke-before-the-us-oil-industry-buckles/.

Famiglietti, J.S. 2014. The Global Groundwater Crisis. *Nature Climate Change* 4(11): 945–948. doi:10.1038/nclimate2425.

Farmer, J., M. Doyne, C. Gallegati, A. Hommes, P. Kirman, S. Ormerod, A. Sanchez Cincotti, and D. Helbing. 2012. A Complex Systems Approach to Constructing Better Models for Managing Financial Markets and the Economy. *The European Physical Journal Special Topics* 214(1): 295–324. doi:10.1140/epjst/e2012-01696-9.

Feely, Richard, Christopher L. Sabine, and Victoria J. Fabry. 2006. Carbon Dioxide and our Ocean Legacy. Pew Trust. http://www.pmel.noaa.gov/pubs/PDF/feel2899/feel2899.pdf.

Foster, John Bellamy, Brett Clark, and Richard York. 2010. *The Ecological Rift: Capitalism's War on the Earth*. New York: NYU Press.

Fournier, Valérie. 2008. Escaping from the Economy: The Politics of Degrowth. *International Journal of Sociology and Social Policy* 28(11/12): 528–545. doi:10.1108/01443330810915233.

Francis. 2014. Boko Haram, Al Shabaab and Al Qaeda 2.0—Islamic Extremism in Africa. *Humanosphere*. May 7. http://www.humanosphere.org/world-politics/2014/05/boko-haram-al-shabaab-and-al-qaeda-2-0-islamic-extremism-in-africa/.

Friedman, Thomas L. 2013. The Scary Hidden Stressor. *The New York Times*, March 2. http://www.nytimes.com/2013/03/03/opinion/sunday/friedman-the-scary-hidden-stressor.html.

Fritz, Martin, and Max Koch. 2014. Potentials for Prosperity without Growth: Ecological Sustainability, Social Inclusion and the Quality of Life in 38 Countries. *Ecological Economics* 108: 191–199. doi:10.1016/j.ecolecon.2014.10.021.

Gagnon, Nathan, Charles A.S. Hall, and Lysle Brinker. 2009. A Preliminary Investigation of Energy Return on Energy Investment for Global Oil and Gas Production. *Energies* 2(3): 490–503. doi:10.3390/en20300490.

García-Olivares, Antonio, and Joaquim Ballabrera-Poy. 2015. Energy and Mineral Peaks, and a Future Steady State Economy. *Technological Forecasting and Social Change* 90, Part B (January): 587–598. doi:10.1016/j.techfore.2014.02.013.

Ghafar, Adel Abdel. 2015. Egypt's New Gas Discovery: Opportunities and Challenges | Brookings Institution. *Brookings*. September 10. https://www.brookings.edu/opinions/egypts-new-gas-discovery-opportunities-and-challenges/.

Guilford, Megan C., Charles A.S. Hall, Peter O'Connor, and Cutler J. Cleveland. 2011. A New Long Term Assessment of Energy Return on Investment (EROI) for U.S. Oil and Gas Discovery and Production. *Sustainability* 3(10): 1866–1887. doi:10.3390/su3101866.

Gülen, Gürcan, John Browning, Svetlana Ikonnikova, and Scott W. Tinker. 2013. Well Economics Across Ten Tiers in Low and High Btu (British Thermal Unit) Areas, Barnett Shale, Texas. *Energy* 60: 302–315. doi:10.1016/j.energy.2013.07.041.

Hall, Charles A. S., and Kent A. Klitgaard. 2012. *Energy and the Wealth of Nations*. New York, NY: Springer New York. http://link.springer.com/10.1007/978-1-4419-9398-4.

Hall, Charles A.S., Cutler J. Cleveland, and Robert K. Kaufmann. 1992. *Energy and Resource Quality: The Ecology of the Economic Process*. Niwot, CO: University Press of Colorado.

Hall, Charles A.S., Jessica G. Lambert, and Stephen B. Balogh. 2014. EROI of Different Fuels and the Implications for Society. *Energy Policy* 64: 141–152. doi:10.1016/j.enpol.2013.05.049.

Hallock Jr., John L., Wei Wu, Charles A.S. Hall, and Michael Jefferson. 2014. Forecasting the Limits to the Availability and Diversity of Global Conventional Oil Supply: Validation. *Energy* 64: 130–153. doi:10.1016/j.energy.2013.10.075.

Ho, Mae-Wan. 1999. Are Economic Systems Like Organisms? In *Sociobiology and Bioeconomics*, ed. Peter Koslowski, 237–258. Studies in Economic Ethics and Philosophy. Berlin: Springer. http://link.springer.com/chapter/10.1007/978-3-662-03825-3_12.

Holling, C.S. 2001. Understanding the Complexity of Economic, Ecological, and Social Systems. *Ecosystems* 4(5): 390–405. doi:10.1007/s10021-001-0101-5.

Holthaus, Eric. 2014. Hot Zone. *Slate*, June 27. http://www.slate.com/articles/technology/future_tense/2014/06/isis_water_scarcity_is_climate_change_destabilizing_iraq.single.html.

Homer-Dixon, Thomas. 2011. *Carbon Shift: How Peak Oil and the Climate Crisis Will Change Canada (and Our Lives)*. Toronto: Random House of Canada.

Hook, Leslie. 2013. China's Appetite for Food Imports to Fuel Agribusiness M&A. *Financial Times*, June 6. http://www.ft.com/cms/s/0/91a2f1ea-cdce-11e2-8313-00144feab7de.html?siteedition=uk#axzz4HZnyd0IC.

Hughes, J. David. 2013. Energy: A Reality Check on the Shale Revolution. *Nature* 494(7437): 307–308. doi:10.1038/494307a.

ICEF. 2016. Growing Chinese Middle Class Projected to Spend Heavily on Education through 2030. ICEF Monitor. http://monitor.icef.com/2016/04/growing-chinese-middle-class-projected-spend-heavily-education-2030/.

IEA. 2009. *World Energy Outlook*. Washington, DC: International Energy Agency.

———. 2015. India Energy Outlook. World Energy Outlook Special Report. International Energy Agency. https://www.iea.org/publications/freepublications/publication/india-energy-outlook-2015.html.

Inman, Mason. 2014. Natural Gas: The Fracking Fallacy. *Nature* 516(7529): 28–30. doi:10.1038/516028a.

IRIN. 2008. Bread Subsidies Under Threat as Drought Hits Wheat Production. *IRIN*. June 30. http://www.irinnews.org/feature/2008/06/30/bread-subsidies-under-threat-drought-hits-wheat-production.

———. 2010. Growing Protests over Water Shortages. *IRIN*. July 27. http://www.irinnews.org/news/2010/07/27/growing-protests-over-water-shortages.

———. 2012. Time Running Out for Solution to Water Crisis. *IRIN*. August 13. http://www.irinnews.org/analysis/2012/08/13/time-running-out-solution-water-crisis.

Jackson, Tim. 2009. *Prosperity Without Growth: Economics for a Finite Planet*. London: Earthscan.

Jackson, Peter M., and Leta K. Smith. 2014. Exploring the Undulating Plateau: The Future of Global Oil Supply. *Philosophical Transactions of the Royal Society of London A: Mathematical, Physical and Engineering Sciences* 372(2006): 20120491. doi:10.1098/rsta.2012.0491.

Jancovici, Jean-Marc. 2013. A Couple of Thoughts in the Energy Transition. Manicore. http://www.manicore.com/anglais/documentation_a/transition_energy.html.

Jefferson, Michael. 2016. A Global Energy Assessment. *Wiley Interdisciplinary Reviews: Energy and Environment* 5(1): 7–15. doi:10.1002/wene.179.

Johanisova, Nadia, and Stephan Wolf. 2012. Economic Democracy: A Path for the Future? *Futures*, Special Issue: Politics, Democracy and Degrowth, 44(6): 562–570. doi:10.1016/j.futures.2012.03.017.

Johnstone, Sarah, and Jeffrey Mazo. 2011. Global Warming and the Arab Spring. *Survival* 53(2): 11–17. doi:10.1080/00396338.2011.571006.

Kaminska, Izabella. 2014. Energy Is Gradually Decoupling from Economic Growth. *FT Alphaville*, January 17. http://ftalphaville.ft.com/2014/01/17/1745542/energy-is-gradually-decoupling-from-economic-growth/.

Katusa, Marin. 2016. How to Pocket Extraordinary Profits from Unconventional Oil. Casey Energy Report.

Kavanagh, Jennifer. 2013. Do U.S. Military Interventions Occur in Clusters? Product Page. http://www.rand.org/pubs/research_briefs/RB9718.html.

Kelley, Colin P., Shahrzad Mohtadi, Mark A. Cane, Richard Seager, and Yochanan Kushnir. 2015. Climate Change in the Fertile Crescent and Implications of the Recent Syrian Drought. *Proceedings of the National Academy of Sciences* 112(11): 3241–3246. doi:10.1073/pnas.1421533112.

King, Carey W. 2015. Comparing World Economic and Net Energy Metrics, Part 3: Macroeconomic Historical and Future Perspectives. *Energies* 8(11): 12997–12920. doi:10.3390/en81112348.

King, Carey W., John P. Maxwell, and Alyssa Donovan. 2015a. Comparing World Economic and Net Energy Metrics, Part 1: Single Technology and Commodity Perspective. *Energies* 8(11): 12949–12974. doi:10.3390/en81112346.

———. 2015b. Comparing World Economic and Net Energy Metrics, Part 2: Total Economy Expenditure Perspective. *Energies* 8(11): 12975–12996. doi:10.3390/en81112347.

Kirkpatrick, David D. 2013a. Egypt, Short of Money, Sees Crisis on Food and Gas. *The New York Times*, March 30. http://www.nytimes.com/2013/03/31/world/middleeast/egypt-short-of-money-sees-crisis-on-food-and-gas.html.

———. 2013b. Egypt, Short of Money, Sees Crisis on Food and Gas. *The New York Times*, March 30. http://www.nytimes.com/2013/03/31/world/middleeast/egypt-short-of-money-sees-crisis-on-food-and-gas.html.

Klump, Edward, and Jim Polson. 2016. Shale-Gas Skeptic's Supply Doubts Draw Wrath of Devon. *Bloomberg.com*. Accessed July 11. http://www.bloomberg.com/news/articles/2009-11-17/shalegas-skeptics-supply-doubts-draw-wrath-of-devon-energy.

Kothari, Ashish. 2014. Degrowth and Radical Ecological Democracy: A View from the South—Blog Postwachstum. *Postwatchstum, Wuppertal Institute*. June 27. http://www.postwachstum.de/degrowth-and-radical-ecological-democracy-a-view-from-the-south-20140627.

Kundu, Tadit. 2016. Nearly Half of Indians Survived on Less than Rs38 a Day in 2011–2012. http://www.livemint.com/. April 20. http://www.livemint.com/Opinion/l1gVncveq4EYEn-2zuzX4FL/Nearly-half-of-Indians-survived-on-less-than-Rs38-a-day-in-2.html.

Lagi, Marco, Karla Z. Bertrand, and Yaneer Bar-Yam. 2011. The Food Crises and Political Instability in North Africa and the Middle East. *arXiv:1108.2455 [physics]*, August. http://arxiv.org/abs/1108.2455.

Lazenby, Henry. 2016. Opec Believed to Overstate Oil Reserves by 70%, Reserves Depleted Sooner. *Mining Weekly*. Accessed August 22. http://www.miningweekly.com/article/opec-believed-to-overstate-oil-reserves-by-70-reserves-depleted-sooner-2012-10-04.

Lelieveld, J., Y. Proestos, P. Hadjinicolaou, M. Tanarhte, E. Tyrlis, and G. Zittis. 2016. Strongly Increasing Heat Extremes in the Middle East and North Africa (MENA) in the 21st Century. *Climatic Change* 137(1–2): 245–260. doi:10.1007/s10584-016-1665-6.

LePoire, David, and Argonne National Laboratory, Argonne, IL, USA. 2015. Interpreting 'big History' as Complex Adaptive System Dynamics with Nested Logistic Transitions in Energy Flow and Organization—Emergence: Complexity and Organization. *Emergence*, March. https://journal.emergentpublications.com/article/interpreting-big-history-as-complex-adaptive-system-dynamics-with-nested-logistic-transitions-in-energy-flow-and-organization/.

Lesk, Corey, Pedram Rowhani, and Navin Ramankutty. 2016. Influence of Extreme Weather Disasters on Global Crop Production. *Nature* 529(7584): 84–87. doi:10.1038/nature16467.

Li, Minqi. 2014. *Peak Oil, Climate Change, and the Limits to China's Economic Growth*. New York: Routledge.

MacDonald, Gregor. 2010. Think OPEC Exports Won't Decline? You're Living In A Dreamworld. *Business Insider*. August 14. http://www.businessinsider.com/think-opec-exports-wont-decline-youre-living-in-a-dreamworld-2010-8.

Matsumoto, Ken'ichi, and Vlasios Voudouris. 2014. Potential Impact of Unconventional Oil Resources on Major Oil-Producing Countries: Scenario Analysis with the ACEGES Model. *Natural Resources Research* 24(1): 107–119. doi:10.1007/s11053-014-9246-8.

Mawry, Yousef. 2015. Yemen Fuel Crisis Ignites Street Riots. *Middle East Eye*. February 12. http://www.middleeasteye.net/news/yemen-fuel-crises-ignites-ongoing-street-riots-393941730.

May, Robert M., Simon A. Levin, and George Sugihara. 2008. Complex Systems: Ecology for Bankers. *Nature* 451(7181): 893–895. doi:10.1038/451893a.

Mayah, Emmanuel. 2012. Climate Change Fuels Nigeria Terrorism. *Africa Review*. February 24. http://www.africareview.com/news/Climate-change-fuels-Nigeria-terrorism/979180-1334472-4m5dlu/index.html.

McGlade, Christophe, Jamie Speirs, and Steve Sorrell. 2013. Unconventional Gas—A Review of Regional and Global Resource Estimates. *Energy* 55: 571–584. doi:10.1016/j.energy.2013.01.048.

Meighan, Brendan. 2016. Egypt's Natural Gas Crisis. *Carnegie Endowment for International Peace*. January. http://carnegieendowment.org/sada/62534.

Moeller, Devin, and David Murphy. 2016. Net Energy Analysis of Gas Production from the Marcellus Shale. *BioPhysical Economics and Resource Quality* 1(1): 1–13. doi:10.1007/s41247-016-0006-8.

Mohr, Steve. 2010. *Projection of World Fossil Fuel Production with Supply and Demand Interactions*. Callaghan: University of Newcastle.

Mohr, S.H., and G.M. Evans. 2009. Forecasting Coal Production until 2100. *Fuel* 88(11): 2059–2067. doi:10.1016/j.fuel.2009.01.032.

———. 2010. Long Term Prediction of Unconventional Oil Production. *Energy Policy* 38(1): 265–276. doi:10.1016/j.enpol.2009.09.015.

Mohr, S.H., J. Wang, G. Ellem, J. Ward, and D. Giurco. 2015. Projection of World Fossil Fuels by Country. *Fuel* 141: 120–135. doi:10.1016/j.fuel.2014.10.030.

Mora, Camilo, Abby G. Frazier, Ryan J. Longman, Rachel S. Dacks, Maya M. Walton, Eric J. Tong, Joseph J. Sanchez, et al. 2013a. The Projected Timing of Climate Departure from Recent Variability. *Nature* 502(7470): 183–187. doi:10.1038/nature12540.

Mora, Camilo, Chih-Lin Wei, Audrey Rollo, Teresa Amaro, Amy R. Baco, David Billett, Laurent Bopp, et al. 2013b. Biotic and Human Vulnerability to Projected Changes in Ocean Biogeochemistry over the 21st Century. *PLOS Biol* 11(10): e1001682. doi:10.1371/journal.pbio.1001682.

Morgan, Geoffrey. 2016. Average Oil Production to Decline This Year, Grow More Slowly in the Future: CAPP. *Financial Post*, June 23. http://business.financialpost.com/news/energy/capp-cuts-canadian-oil-output-forecast-to-4-9-million-barrels-a-day-by-2030?__lsa=2a41-2f74.

Morrissey, John. 2016. US Central Command and Liberal Imperial Reach: Shaping the Central Region for the 21st Century. *The Geographical Journal* 182(1): 15–26. doi:10.1111/geoj.12118.

Murphy, David J. 2014. The Implications of the Declining Energy Return on Investment of Oil Production. *Philosophical Transactions of the Royal Society of London A: Mathematical, Physical and Engineering Sciences* 372(2006): 20130126. doi:10.1098/rsta.2013.0126.

Murphy, David J., and Charles A.S. Hall. 2011. Energy Return on Investment, Peak Oil, and the End of Economic Growth. *Annals of the New York Academy of Sciences* 1219(1): 52–72. doi:10.1111/j.1749-6632.2010.05940.x.

Nandi, Sanjib Kumar. 2014. A Study on Hubbert Peak of India's Coal: A System Dynamics Approach. *International Journal of Scientific & Engineering Research* 9(2). http://www.academia.edu/9744358/A_Study_on_Hubbert_Peak_of_Indias_Coal_A_System_Dynamics_Approach.

Nekola, Jeffrey C., Craig D. Allen, James H. Brown, Joseph R. Burger, Ana D. Davidson, Trevor S. Fristoe, Marcus J. Hamilton, et al. 2013. The Malthusian–Darwinian Dynamic and the Trajectory of Civilization. *Trends in Ecology & Evolution* 28(3): 127–130. doi:10.1016/j.tree.2012.12.001.

OBG. 2016. New Discoveries for Egyptian Oil Producers. *Oxford Business Group*. January 27. http://www.oxfordbusinessgroup.com/overview/fresh-ideas-new-discoveries-have-oil-producers-optimistic-about-future.

Odhiambo, George O. 2016. Water Scarcity in the Arabian Peninsula and Socio-Economic Implications. *Applied Water Science*, June, 1–14. doi:10.1007/s13201-016-0440-1.

Odum, Howard Thomas. 1994. *Ecological and General Systems: An Introduction to Systems Ecology*. Niwot, CO: University Press of Colorado.

Omisore, Bolanle. 2014. Nigerians Face Fuel Shortages In the Shadow of Plenty. *National Geographic News*. April 11. http://news.nationalgeographic.com/news/energy/2014/04/140411-nigeria-fuel-shortage-oil/.

Onyia, Chukwuma. 2015. Climate Change and Conflict in Nigeria: The Boko Haram Challenge. *American International Journal of Social Science* 4(2). http://www.aijssnet.com/journal/index/329.

Owen, Nick A., Oliver R. Inderwildi, and David A. King. 2010. The Status of Conventional World Oil reserves—Hype or Cause for Concern? *Energy Policy* 38(8): 4743–4749. doi:10.1016/j.enpol.2010.02.026.

Patrick, Roger. 2015. When the Well Runs Dry: The Slow Train Wreck of Global Water Scarcity. *Journal—American Water Works Association* 107: 65–76. doi:10.5942/jawwa.2015.107.0042.

Patzek, Tad W., Frank Male, and Michael Marder. 2013. Gas Production in the Barnett Shale Obeys a Simple Scaling Theory. *Proceedings of the National Academy of Sciences* 110(49): 19731–19736. doi:10.1073/pnas.1313380110.

Pearce, Joshua M. 2008. Thermodynamic Limitations to Nuclear Energy Deployment as a Greenhouse Gas Mitigation Technology. *International Journal of Nuclear Governance, Economy and Ecology* 2(1): 113. doi:10.1504/IJNGEE.2008.017358.

Peel, Michael. 2013. Subsidies 'Distort' Saudi Arabia Economy Says Economy Minister. *Financial Times*. May 7. http://www.ft.com/cms/s/0/f474cf28-b717-11e2-841e-00144feabdc0.html.

Phys.org. 2016. Minority Rules: Scientists Discover Tipping Point for the Spread of Ideas. Accessed August 21. http://phys.org/news/2011-07-minority-scientists-ideas.html.

Pichler, Franz. 1999. Modeling Complex Systems by Multi-Agent Holarchies. In *Computer Aided Systems Theory—EUROCAST'99*, ed. Peter Kopacek, Roberto Moreno-Díaz, and Franz Pichler, 154–168. Lecture Notes in Computer Science 1798. Springer Berlin Heidelberg. http://link.springer.com/chapter/10.1007/10720123_14.

Pierce, Charles P. 2016. What Happens When the American Southwest Runs Out of Water? *Esquire*. June 1. http://www.esquire.com/news-politics/politics/news/a45398/southwest-desert-water-drought/.

Pracha, Ali S., and Timothy A. Volk. 2011. An Edible Energy Return on Investment (EEROI) Analysis of Wheat and Rice in Pakistan. *Sustainability* 3(12): 2358–2391. doi:10.3390/su3122358.

Pritchard, Bill. 2016. The Impacts of Climate Change for Food and Nutrition Security: Issues for India. In *Climate Change Challenge (3C) and Social-Economic-Ecological Interface-Building*. Environmental Science and Engineering. Springer.

Pueyo, Salvador. 2014. Ecological Econophysics for Degrowth. *Sustainability* 6(6): 3431–3483. doi:10.3390/su6063431.

Qaed, Samar. 2014. Expanding Too Quickly? *Yemen Times*. February 25. http://www.yementimes.com/en/1758/report/3522/Expanding-too-quickly.htm.

Qi, Ye, Nicholas Stern, Tong Wu, Jiaqi Lu, and Fergus Green. 2016. China's Post-Coal Growth. *Nature Geoscience* 9. doi:10.1038/ngeo2777.

Reganold, John P., and Jonathan M. Wachter. 2016. Organic Agriculture in the Twenty-First Century. *Nature Plants* 2(2): 15221. doi:10.1038/nplants.2015.221.

Rioux, Sébastien, and Frédérick Guillaume Dufour. 2008. La sociologie historique de la théorie des relations sociales de propriété. *Actuel Marx* 43(1): 126. doi:10.3917/amx.043.0126.

RiskMetrics Group. 2010. Canada's Oil Sands: Shrinking Window of Opportunity. Ceres, Inc. http://www.ceres.org/resources/reports/oil-sands-2010.

Rockström, Johan, Will Steffen, Kevin Noone, Persson Åsa, F. Stuart Chapin, Eric F. Lambin, Timothy M. Lenton, et al. 2009. A Safe Operating Space for Humanity. *Nature* 461(7263): 472–475. doi:10.1038/461472a.

Ross, John, and Adam P. Arkin. 2009. Complex Systems: From Chemistry to Systems Biology. *Proceedings of the National Academy of Sciences* 106(16): 6433–6434. doi:10.1073/pnas.0903406106.

Salameh, M. G. 2012. Impact of US Shale Oil Revolution on the Global Oil Market, the Price of Oil & Peak Oil. http://www.rcem.eu/research/rcemworkingpapers/impact-of-us-shale-oil--revolution-on-the-global-oil-market,-the-price-of-oil-peak-oil.aspx.

Saleh, Hebah. 2013. Egypt Weighs Burden of IMF Austerity. *Financial Times*. March 11. http://www.ft.com/cms/s/0/464a9350-8a6d-11e2-bf79-00144feabdc0.html.

Sanders, Jim. 2013. The Hidden Force behind Islamic Militancy in Nigeria? Climate Change. *The Christian Science Monitor*. July 8. http://m.csmonitor.com/World/Africa/Africa-Monitor/2013/0708/The-hidden-force-behind-Islamic-militancy-in-Nigeria-Climate-change.

Sands, Phil. 2011. Population Surge in Syria Hampers Country's Progress | The National. *The National*, March 6. http://www.thenational.ae/news/world/middle-east/population-surge-in-syria-hampers-countrys-progress.

Sarant, Louise. 2013. Climate Change and Water Mismanagement Parch Egypt | Egypt Independent. *Egypt Independent*. February 26. http://www.egyptindependent.com/news/climate-change-and-water-mismanagement-parch-egypt.

Sayne, Aaron. 2011. Climate Change Adaptation and Conflict in Nigeria. Special Report. United States Institute of Peace. http://www.usip.org/publications/climate-change-adaptation-and-conflict-in-nigeria.

Schneider, E.D., and J.J. Kay. 1994. Life as a Manifestation of the Second Law of Thermodynamics. *Mathematical and Computer Modelling* 19(6): 25–48. doi:10.1016/0895-7177(94)90188-0.

Schneider, François, Giorgos Kallis, and Joan Martinez-Alier. 2010. Crisis or Opportunity? Economic Degrowth for Social Equity and Ecological Sustainability. Introduction to This Special Issue. *Journal of Cleaner Production*, Growth, Recession or Degrowth for Sustainability and Equity? 18(6): 511–518. doi:10.1016/j.jclepro.2010.01.014.

Schrodinger, Erwin. 1944. *What Is Life?* http://whatislife.stanford.edu/LoCo_files/What-is-Life.pdf.

Schwartzman, David, and Peter Schwartzman. 2013. A Rapid Solar Transition Is Not Only Possible, It Is Imperative! *African Journal of Science, Technology. Innovation and Development* 5(4): 297–302. doi:10.1080/20421338.2013.809260.

Shahine, Alaa. 2016. Egypt Had FDI Outflows of $482.7 Million in 2011. *Bloomberg.com*. Accessed August 16. http://www.bloomberg.com/news/articles/2012-03-25/egypt-had-fdi-outflows-of-482-7-million-in-2011-correct-.

Shaw, Martin. 2005. Risk-Transfer Militarism and the Legitimacy of War after Iraq. In *September 11, 2001: A Turning-Point in International and Domestic Law?* ed. Paul Eden and T. O'Donnell. Transnational Publishers. http://sro.sussex.ac.uk/12462/.

Simms, Andrew. 2008. The Poverty Myth. *New Scientist* 200(2678): 49. doi:10.1016/S0262-4079(08)62641-X.

Smith-Nonini, Sandy. 2016. The Role of Corporate Oil and Energy Debt in Creating the Neoliberal Era. *Economic Anthropology* 3(1): 57–67. doi:10.1002/sea2.12044.

Söderbergh, Bengt, Fredrik Robelius, and Kjell Aleklett. 2007. A Crash Programme Scenario for the Canadian Oil Sands Industry. *Energy Policy* 35(3): 1931–1947. doi:10.1016/j.enpol.2006.06.007.

Steffen, Will, Katherine Richardson, Johan Rockström, Sarah E. Cornell, Ingo Fetzer, Elena M. Bennett, R. Biggs, et al. 2015. Planetary Boundaries: Guiding Human Development on a Changing Planet. *Science*, January, 1259855. doi:10.1126/science.1259855.

Stewart, Ian. 2015. Debt-Driven Growth, Where Is the Limit? *Deloitte: Monday Briefing*. February 2. http://blogs.deloitte.co.uk/mondaybriefing/2015/02/debt-driven-growth-where-is-the-limit.html.

Stokes, Doug, and Sam Raphael. 2010. *Global Energy Security and American Hegemony*. Baltimore: JHU Press.

Stott, Peter. 2016. How Climate Change Affects Extreme Weather Events. *Science* 352(6293): 1517–1518. doi:10.1126/science.aaf7271.

Street, 1615 L., NW, Suite 800 Washington, and DC 20036 202 419 4300 | Main 202 419 4349 | Fax 202 419 4372 | Media Inquiries. 2014. Attitudes about Aging: A Global Perspective. *Pew Research Center's Global Attitudes Project*. January 30. http://www.pewglobal.org/2014/01/30/attitudes-about-aging-a-global-perspective/.

Taha, Sharif. 2014. Kingdom Imports 80% of Food Products. *Arab News*. April 20. http://www.arabnews.com/news/558271.

Tainter, Joseph. 1990. *The Collapse of Complex Societies*. Cambridge: Cambridge University Press.

Tao, Fulu, Masayuki Yokozawa, Yousay Hayashi, and Erda Lin. 2003. Future Climate Change, the Agricultural Water Cycle, and Agricultural Production in China. *Agriculture, Ecosystems & Environment* 95(1): 203–215. doi:10.1016/S0167-8809(02)00093-2.

TE. 2016. Egypt Government Debt to GDP | 2002-2016 | Data | Chart | Calendar. *Trading Economics*. http://www.tradingeconomics.com/egypt/government-debt-to-gdp.

Terzis, George, and Robert Arp, eds. 2011. *Information and Living Systems: Philosophical and Scientific Perspectives*. MIT Press. http://www.jstor.org/stable/j.ctt5hhhvb.

Thevard, Benoit. 2012. Europe Facing Peak Oil. Momentum Institute/Greens-EFA Group in European Parliament. http://www.greens-efa.eu/fileadmin/dam/Documents/Publications/PIC%20petrolier_EN_lowres.pdf.

Timms, Matt. 2016. Resource Mismanagement Has Led to a Critical Water Shortage in Asia. *World Finance*, July 21. http://www.worldfinance.com/infrastructure-investment/government-policy/resource-mismanagement-has-led-to-a-critical-water-shortage-in-asia.

Tong, Shilu et al. 2016. Climate Change, Food, Water and Population Health in China. *Bulletin of the World Health Organization*, July. http://www.who.int/bulletin/online_first/BLT.15.167031.pdf?ua=1.

Tranum, Sam. 2013. *Powerless: India's Energy Shortage and Its Impact*. India: Sage.

Trendberth, Kevin, Jerry Meehl, Jeff Masters, and Richard Somerville. 2012. Heat Waves and Climate Change. https://www.climatecommunication.org/wp-content/uploads/2012/06/Heat_Waves_and_Climate_Change.pdf.

Tverberg, Gail. 2016. China: Is Peak Coal Part of Its Problem? *Our Finite World*. June 20. https://ourfiniteworld.com/2016/06/20/china-is-peak-coal-part-of-its-problem/.

UN. 2015. World Population Prospects. United Nations Department of Economic & Social Affairs, Population Division.

UN News Center, United Nations News Service. 2012. UN News—Despite End-of-Year Decline, 2011 Food Prices Highest on Record—UN. *UN News Service Section*. January 12. http://www.un.org/apps/news/story.asp?NewsID=40925#.V7L65Vcqbdm.

Victor, Peter. 2010. Questioning Economic Growth. *Nature* 468(7322): 370–371. doi:10.1038/468370a.

Vyas, Kejal, and Timothy Puko. 2016. Venezuela Oil Production Drops Sharply in May. *Wall Street Journal*, June 14, sec. World. http://www.wsj.com/articles/venezuela-oil-production-drops-sharply-in-may-1465868354.

Wang, Jinxia, Robert Mendelsohn, Ariel Dinar, Jikun Huang, Scott Rozelle, and Lijuan Zhang. 2009. The Impact of Climate Change on China's Agriculture. *Agricultural Economics* 40(3): 323–337. doi:10.1111/j.1574-0862.2009.00379.x.

Wang, Ke, Lianyong Feng, Jianliang Wang, Yi Xiong, and Gail E. Tverberg. 2016. An Oil Production Forecast for China Considering Economic Limits. *Energy* 113: 586–596. doi:10.1016/j.energy.2016.07.051.

Weijermars, Ruud. 2013. Economic Appraisal of Shale Gas Plays in Continental Europe. *Applied Energy* 106: 100–115. doi:10.1016/j.apenergy.2013.01.025.

Wiedmann, Thomas O., Heinz Schandl, Manfred Lenzen, Daniel Moran, Sangwon Suh, James West, and Keiichiro Kanemoto. 2015. The Material Footprint of Nations. *Proceedings of the National Academy of Sciences* 112(20): 6271–6676. doi:10.1073/pnas.1220362110.

Wilkinson, Henry. 2016. Political Violence Contagion: A Framework for Understanding the Emergence and Spread of Civil Unrest. Lloyd's. http://www.lloyds.com/~/media/files/news%20and%20insight/risk%20insight/2016/political%20violence%20contagion.pdf.

Williams, Selina, and Bradley Olson. 2016. Big Oil Companies Binge on Debt. *Wall Street Journal*, August 24. http://www.wsj.com/articles/largest-oil-companies-debts-hit-record-high-1472031002.

Wood, Ellen Meiksins. 1981. The Separation of the Economic and the Political in Capitalism. *New Left Review, I* 127: 66–95.

World Bank. 2014. Future Impact of Climate Change Visible Now in Yemen. Text/HTML. *World Bank*. November 24. http://www.worldbank.org/en/news/feature/2014/11/24/future-impact-of-climate-change-visible-now-in-yemen.

Worth, Robert F. 2010. Drought Withers Lush Farmlands in Syria. *The New York Times*, October 13. http://www.nytimes.com/2010/10/14/world/middleeast/14syria.html.

Yaritani, Hiroaki, and Jun Matsushima. 2014. Analysis of the Energy Balance of Shale Gas Development. *Energies* 7(4): 2207–2227. doi:10.3390/en7042207.

Index

N.M. Ahmed, *Failing States, Collapsing Systems*, SpringerBriefs in Energy,
DOI 10.1007/978-3-319-47816-6

CPSIA information can be obtained
at www.ICGtesting.com
Printed in the USA
LVHW060026190319
611108LV00002B/2/P